Medal of Honor Heroes: World War II

15 True Stories of Courage and Sacrifice for Ages 8–12

Part of the Medal of Honor Heroes series from Good Buddy Books

Published by Good Buddy Books

goodbuddybooks.com

ISBN: 978-1-968726-35-5

First edition

Cover design by Good Buddy Books
Interior design by Good Buddy Books

This book is a work of narrative nonfiction based on historical sources believed to be accurate at the time of publication.

This book is intended for educational and informational purposes.

Table of Contents

Medal of Honor Heroes Foreword

What Is the Medal of Honor?

Throughout history, many soldiers have shown bravery in battle. But only a very small number have performed acts of courage so extraordinary that the United States awards them the nation's highest military honor.

That award is called the **Medal of Honor**.

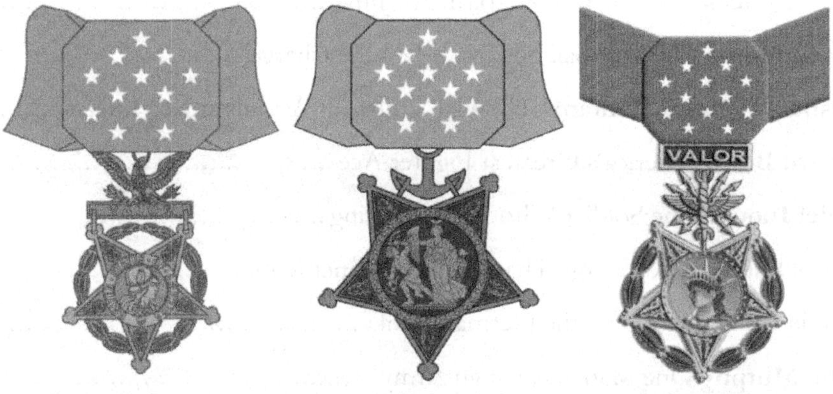

ARMY **NAVY** **AIR FORCE**

Each branch has a different design for the Medal of Honor

The Medal of Honor is given to members of the United States Armed Forces who go **"above and beyond the call of duty."** In other words, these are people who risk their lives to protect others, even when the danger is extreme.

Since the medal was first created during the Civil War in the 1860s, it has been awarded to just over **3,500 service members**. Millions of Americans have served in the military during that time. Only a tiny fraction have received the Medal of Honor.

Each medal tells the story of a remarkable moment.

Sometimes a soldier charged forward to stop an enemy attack. Sometimes a pilot protected fellow airmen in the sky. Sometimes a medic risked his life again and again to rescue wounded soldiers.

Many Medal of Honor recipients never expected recognition for what they did. In fact, many of them simply believed they were doing their duty—helping their fellow soldiers when they needed it most.

The **true stories** in this book are about those moments.

You will meet soldiers who faced impossible choices. Some of them survived the battles they fought. Others gave their lives while protecting their friends.

But every one of them showed a kind of courage that still inspires people today.

These stories also remind us that courage can take many forms. Some heroes fought with rifles. Others saved lives as medics. Some led daring attacks. Others stood their ground when retreat seemed easier.

No matter how their bravery appeared, they all shared something important: they chose to help others even when it meant risking everything.

As you read these chapters, think about the decisions these heroes made.

What would you do in a moment like that?

The stories in this book are not just about battles in the past. They are about character, courage, and the choices people make when others depend on them.

And sometimes, those choices change history.

Desmond Doss

The Medic Who Saved 75 Soldiers

Gunfire crackled across the ridge.

American soldiers scrambled for cover as explosions shook the rocky ground. The steep cliff behind them dropped nearly four hundred feet to the jungle below.

Moments earlier, the men had been advancing across the top of the ridge. Now many of them were wounded. Some lay motionless in the dirt. Others cried out for help.

Most of the soldiers had already retreated down the cliff.

But one man remained.

Private Desmond Doss moved carefully across the battlefield, searching for the wounded. He carried no rifle and wore no weapon on his belt. Instead, his medical bag hung over his shoulder.

Bullets snapped through the air around him.

Doss knelt beside a wounded soldier and quickly wrapped a bandage around the man's leg. Then he dragged him toward the edge of the cliff.

Below them, ropes dangled down the rocky wall.

Working as fast as he could, Doss tied the rope around the soldier's body and began lowering him down the cliff to safety.

When the man reached the bottom, Doss whispered a quick prayer.

Then he turned around and ran back into the gunfire.

A Quiet Boy from Virginia

Long before he became a war hero, Desmond Doss was a quiet young man from the hills of Virginia.

He was born on February 7, 1919, in the city of Lynchburg. Life there was simple. Desmond's father worked as a carpenter, and the family did not have much money. But his home was filled with strong beliefs about faith, kindness, and respect for others.

Those beliefs shaped the way Desmond saw the world.

His family were devoted Seventh-day Adventists, and their faith taught that human life was sacred. One of the Bible's commandments—"Thou shalt not kill"—was especially important to Desmond. As he grew older, he decided that he would never take another person's life.

Friends later described him as quiet, polite, and thoughtful. He was not the kind of person who looked for attention or arguments. Instead, he tried to live by the values he believed were right.

Then the world changed.

On December 7, 1941, Japan attacked the American naval base at Pearl Harbor in Hawaii. The surprise attack killed more than 2,400 Americans and pulled the United States into World War II.

Across the country, young men began joining the military.

Desmond Doss decided he would serve too.

Desmond Doss portrait

He believed his country needed him. But he also believed just as strongly that he could not carry a weapon or kill another human being.

So when he joined the United States Army in 1942, Doss made an unusual choice.

Instead of becoming a rifleman, he trained to become a **combat medic**—a soldier whose job was to treat wounded men and carry them to safety.

Many other soldiers did not understand his decision.

Some thought a man who refused to carry a rifle did not belong in the army at all.

But Desmond Doss believed he could serve his country in another way.

Soon, that belief would be tested on one of the deadliest battlefields of World War II.

War in the Pacific

World War II was fought across much of the world.

American forces battled Germany in Europe and North Africa. At the same time, another fierce war was unfolding across thousands of miles of ocean in the Pacific.

There, the United States and its allies were fighting the Empire of Japan.

The war in the Pacific was unlike anything most American soldiers had experienced before. Battles were often fought on small islands surrounded by jungle, steep cliffs, and rocky beaches. The fighting was intense and dangerous. Soldiers faced enemy fire, hidden bunkers, and difficult terrain.

By 1945, the war had been raging for nearly four years.

American forces had slowly pushed across the Pacific, island by island. Each battle brought them closer to Japan itself.

One of the final and most important steps was the island of **Okinawa**.

Okinawa lies about 350 miles south of Japan's main islands. Military leaders believed that capturing Okinawa would give American forces a place to launch air attacks and prepare for a possible invasion of Japan.

But the Japanese army was determined to defend the island at all costs.

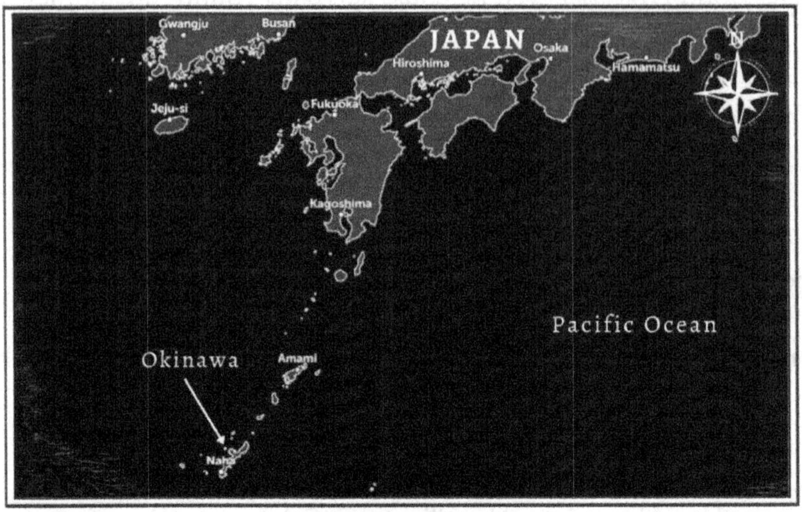

OKINAWA: 1945

When American troops landed on Okinawa in April 1945, they soon discovered that the battle would be long and brutal. Japanese soldiers had built strong defensive positions in caves, tunnels, and bunkers across the island.

One of the most dangerous places on Okinawa was a steep cliff that American soldiers nicknamed **Hacksaw Ridge**.

At the top of that ridge, the Japanese army had built a powerful defensive line.

And it was there that Desmond Doss would soon face the greatest test of his life.

Into the Battle

The battle for Okinawa began on April 1, 1945.

At first, American forces met little resistance on the beaches. But the calm did not last long. Japanese soldiers had prepared their defenses farther inland, where the land rose into steep hills and ridges.

One of those ridges was the towering escarpment the Americans called **Hacksaw Ridge**.

The cliff rose nearly 400 feet above the ground below. The rocky wall was so steep that soldiers could not simply walk up it. Instead, they used cargo nets—large rope ladders—to climb to the top.

Once they reached the ridge, the soldiers of the U.S. Army's **77th Infantry Division** faced a deadly challenge.

Japanese troops had dug deep defensive positions into the rocky ground. Hidden machine guns and artillery covered the open areas of the ridge. As soon as American soldiers moved forward, they often came under heavy fire.

Desmond Doss climbed those cargo nets alongside the other soldiers.

As a combat medic, he carried no rifle. His job was not to attack the enemy, but to move with the infantry and treat wounded soldiers wherever they fell. Doss carried bandages, medical supplies, and morphine in his field bag.

That meant he had to move into the same dangerous areas as the riflemen.

During the first attacks on Hacksaw Ridge, the fighting quickly turned fierce. Explosions shook the ground. Machine gun fire swept across the ridge. Many soldiers were hit as they tried to push forward.

Soon, the battlefield was filled with wounded men who could not escape on their own.

And that was when Desmond Doss began doing something that few soldiers had ever seen before.

One Man Stayed Behind

The fighting on Hacksaw Ridge grew worse as the day went on.

American soldiers pushed forward across the rocky ground, but Japanese defenders hidden in caves and bunkers opened fire again and again. Explosions echoed across the ridge. Machine guns swept the open areas where the Americans tried to advance.

Soon the battlefield was filled with wounded men.

Medics rushed from soldier to soldier, trying to stop bleeding and carry the injured to safety. Desmond Doss moved quickly through the chaos, kneeling beside wounded soldiers and wrapping bandages around shattered arms and legs.

Again and again, he helped carry injured men toward the cargo nets that hung over the edge of the cliff.

But the fighting did not stop.

At one point, Japanese forces launched a powerful counterattack. Gunfire erupted across the ridge, forcing many American soldiers to retreat down the cliff to regroup.

The wounded could not move fast enough to escape.

Some soldiers were too badly hurt to crawl. Others had been knocked unconscious by explosions.

As the American troops withdrew, dozens of wounded men were left behind on the battlefield.

Most soldiers believed the wounded would have to wait until morning.

But Desmond Doss stayed on the ridge.

Moving carefully through the smoke and debris, he searched for injured soldiers one by one.

When he found a man who could not move, Doss treated his wounds as best he could. Then he dragged or carried the soldier toward the edge of the cliff.

At the edge of the ridge, ropes hung down the steep rock wall. Soldiers below were waiting to pull wounded men to safety.

Doss tied a rope around the first wounded soldier and carefully lowered him down the cliff.

When the man reached the bottom, Doss whispered a short prayer.

"Lord, please help me get one more."

Then he turned and ran back into the battlefield.

Again and again, he searched the ridge for wounded soldiers.

One man had been shot in the leg and could not stand. Doss lifted the soldier onto his shoulders and carried him across the rocky ground while bullets cracked through the air nearby.

Another soldier lay unconscious near a shell crater. Doss dragged him across the ridge and lowered him down the cliff the same way.

Each time he sent a wounded man to safety, Doss returned to the battlefield.

Again he whispered the same prayer.

"Lord, help me get one more."

The night grew darker, but the fighting did not fully stop. Japanese soldiers were still nearby, and gunfire sometimes erupted without warning.

Still, Doss kept working.

He crawled across the battlefield searching for anyone still alive. When he found a wounded soldier, he treated the injuries as best he could before dragging or carrying the man to the edge of the ridge.

Then he lowered the soldier down the cliff to waiting hands below.

Hour after hour, he continued the dangerous work.

By the time Doss finally climbed down the ridge himself, he had rescued about **75 wounded soldiers**.

Many of those men would not have survived the night without him.

The medic who refused to carry a weapon had done something extraordinary.

And the soldiers who once doubted him would never forget it.

The Battle Ends

When Desmond Doss finally climbed down from Hacksaw Ridge, the soldiers waiting below could hardly believe what he had done.

One after another, wounded men had been lowered down the cliff through the darkness. Each time the rope moved again, another injured soldier appeared.

By the end of the night, dozens of men had been saved.

Many of the soldiers watching from below knew Doss well. During training, some of them had doubted him. A few had even mocked him for refusing to carry a rifle. They believed a soldier who would not fight had no place on the battlefield.

But the events on Hacksaw Ridge changed their minds forever.

Those same soldiers now understood that the quiet medic had shown a different kind of courage.

Doss continued serving as a combat medic during the battle for Okinawa. The fighting lasted nearly three months and became one of the bloodiest battles of World War II.

Even after the night on Hacksaw Ridge, Doss kept risking his life to treat wounded soldiers.

Later in the battle, he was seriously wounded by a grenade while helping other men. As medics arrived to carry him away, Doss insisted that they first treat another injured soldier nearby.

Only after that soldier was taken care of did Doss allow himself to be moved.

The quiet medic who once faced doubt from his fellow soldiers had become one of the bravest men on the battlefield.

Soon, the entire nation would hear his story.

After the war ended, news of Desmond Doss's bravery spread across the United States.

Soldiers who had fought on Okinawa told the story of the quiet medic who stayed behind on Hacksaw Ridge and rescued dozens of wounded men. Military leaders carefully reviewed what had happened during the battle.

What they discovered was extraordinary.

Desmond Doss, on top of the Maeda Escarpment, Battle of Okinawa

The Medal of Honor

For his actions on Okinawa, Desmond Doss was awarded the **Medal of Honor**, the highest military award given by the United States.

On **October 12, 1945**, Doss traveled to the White House in Washington, D.C., where President **Harry S. Truman** presented him with the medal.

As Truman placed the medal around Doss's neck, he reportedly said that he would rather have that medal than be president of the United States.

The official citation praised Doss for his courage and determination during the battle on Hacksaw Ridge. It described how he remained on the battlefield while others retreated and rescued about **75 wounded soldiers**, lowering them down the steep cliff to safety.

Desmond Doss became the **first conscientious objector in American history to receive the Medal of Honor**.

The quiet medic from Virginia had proven that courage can take many forms.

President Truman awards the Medal of Honor to Desmond Doss

Why His Courage Still Matters

Desmond Doss never fired a weapon in battle.

Yet his bravery saved the lives of dozens of soldiers on Hacksaw Ridge.

Doss believed deeply in protecting others, even when doing so placed him in great danger. While many soldiers fought with rifles, he fought the battle in a different way—by refusing to leave wounded men behind.

His courage did not come from anger or from the desire to defeat an enemy. It came from his belief that every life was valuable.

Think about the moment when Doss stood alone on the ridge. The battle was still dangerous. The wounded soldiers could not escape on their own.

He could have climbed down the cliff and followed the others to safety.

Instead, he chose to go back again and again to rescue more men.

What kind of courage does it take to make a choice like that?

The story of Desmond Doss reminds us that bravery is not always about fighting harder than everyone else.

Sometimes it means standing by your beliefs—and helping others when they need you most.

John Basilone

Holding the Line on Guadalcanal

The jungle around the Marines was dark and loud.

Rain had fallen earlier, leaving the ground muddy and slippery. The thick trees and tangled vines of **Guadalcanal** surrounded the American positions. Somewhere in the darkness, soldiers could hear movement.

Then the shouting began.

Japanese troops were charging toward the Marine lines.

Moments later, gunfire erupted across the battlefield. Machine guns rattled, rifles cracked, and flares burst into the sky, lighting the jungle with flashes of white.

The attack had begun.

At one of the Marine defensive positions stood **Sergeant John Basilone**, a machine-gun squad leader in the **7th Marines, 1st Marine Division**.

His job was to stop the attack.

The Japanese soldiers kept coming through the darkness, charging again and again toward the Marine lines. Basilone's machine guns fired almost nonstop, cutting through the night as his team fought to hold their ground.

But the battle quickly grew desperate.

Enemy fire tore through the jungle. Ammunition ran low. One by one, Basilone's machine guns began to fall silent.

Some guns were damaged. Others ran out of ammunition. Several Marines were wounded.

Soon, only one man was left to keep the defense alive.

Sergeant John Basilone.

And the battle was far from over.

A Marine from New Jersey

Long before the battle on Guadalcanal, John Basilone grew up far from the jungles of the Pacific.

He was born on **November 4, 1916**, in Buffalo, New York, but his family soon moved to **Raritan, New Jersey**, where he spent most of his childhood. Basilone was one of ten children in a hardworking Italian-American family.

Life in Raritan was simple. As a boy, John enjoyed sports and outdoor adventures. Friends remembered him as energetic and determined—someone who liked a challenge and rarely backed down.

When he was still a young man, Basilone decided to join the **United States Army**. He served for several years, including time in the **Philippines**, where he gained experience as a soldier.

It was during this time that fellow soldiers began calling him **"Manila John."** The nickname came from the capital city of the Philippines, where he had spent part of his service.

John Basilone

After finishing his time in the Army, Basilone returned home to New Jersey and worked various jobs. But as war spread across the world in the late 1930s and early 1940s, he felt the pull of military service again.

In **1940**, Basilone joined the **United States Marine Corps**.

The Marines trained hard for combat. Basilone quickly proved himself to be a skilled and dependable soldier, and he was eventually assigned to a **machine-gun section** in the **7th Marines, 1st Marine Division**.

Not long after the United States entered World War II, Basilone and his fellow Marines were sent across the Pacific Ocean.

Their destination was a small island that would soon become one of the most important battlefields of the war.

Guadalcanal.

The Fight for Guadalcanal

In 1942, World War II in the Pacific was still in its early stages.

Just months earlier, Japan had launched surprise attacks across the Pacific Ocean, including the strike on **Pearl Harbor** that brought the United States into the war. Japanese forces quickly captured large areas of territory, spreading across islands and coastlines throughout the region.

American military leaders knew they had to stop that advance.

One of the most important places in that effort was a small island in the **Solomon Islands** called **Guadalcanal.**

GUADALCANAL: 1942

At first glance, Guadalcanal did not seem very important. It was covered in thick jungle, steep hills, and muddy rivers. But the island held something that could change the course of the war.

The Japanese were building an **airfield** there.

If completed, that airfield would allow Japanese aircraft to threaten American ships moving between the United States and Australia. Controlling the skies above Guadalcanal could give Japan a major advantage in the Pacific.

In **August 1942**, American Marines launched an invasion of the island. They captured the unfinished airfield and began using it themselves. The Marines renamed it **Henderson Field**.

But the Japanese were determined to take it back.

Again and again, Japanese forces attacked the American positions around the airfield. Some battles took place during the day, but many of the fiercest fights happened at night, when Japanese troops tried to break through the Marine defenses.

One of those attacks would become one of the most intense battles of the entire campaign.

And Sergeant John Basilone was right in the middle of it.

Into the Night Battle

By October 1942, the fighting on Guadalcanal had already been going on for weeks.

American Marines held the ground around **Henderson Field**, but Japanese forces continued to attack again and again. Both sides knew that controlling the airfield could decide the battle for the island.

The Marines dug defensive positions along the jungle ridges and hills surrounding the airfield. These lines of defense stretched across muddy ground and thick vegetation where visibility was often poor.

John Basilone's machine-gun section was positioned along one of these defensive lines.

Machine guns were among the most important weapons in the Marines' defense. A single gun could fire hundreds of rounds in minutes, stopping enemy soldiers from charging across open ground. Because of this, machine-gun teams were often placed at key points in the defensive line.

Basilone commanded **two sections of heavy machine guns**.

Each gun required several Marines to operate it. One Marine fired the weapon while others carried ammunition, changed barrels, and helped keep the gun working during the heat of battle.

When the Japanese attack began on the night of **October 24, 1942**, Basilone and his Marines were ready at their positions.

Soon, the sounds of movement began echoing through the jungle.

Japanese troops were approaching.

Then the first shots rang out.

Machine guns thundered across the dark hills as waves of Japanese soldiers charged toward the Marine lines. The attack grew more intense with every passing minute.

Basilone's guns fired continuously, cutting through the darkness as the Marines fought to hold their ground.

But the battle was only beginning.

And before the night was over, Sergeant John Basilone would be fighting almost entirely on his own.

Holding the Line

The Japanese attack grew stronger as the night went on.

Wave after wave of soldiers rushed through the jungle toward the Marine defensive lines. Some charged straight toward the machine-gun positions while others tried to move through the darkness to attack from different directions.

Basilone's machine guns fired almost nonstop.

The loud rattling of the guns echoed across the hills as the Marines tried to stop the advancing enemy. In the flashes of gunfire and exploding grenades, the battlefield lit up again and again.

At first, Basilone's entire team was fighting together.

But the battle quickly turned chaotic.

Enemy fire slammed into the Marine positions. Several gunners were wounded, and one of Basilone's machine guns was knocked out of action. The remaining guns continued firing, but the pressure from the attacking Japanese soldiers kept building.

Soon, Basilone realized something terrible.

Ammunition was running low.

Without ammunition, the machine guns would fall silent—and if that happened, the Japanese soldiers could break through the Marine lines.

Basilone made a decision.

Leaving the safety of his position, he ran through the dark jungle to reach a supply area behind the lines. Enemy soldiers were still moving through the trees, and gunfire cracked through the night as he pushed forward.

At the supply area, Basilone found boxes of ammunition.

Then he began carrying them back.

One by one, he hauled the heavy boxes through the jungle under enemy fire, bringing desperately needed ammunition to the machine-gun crews.

But the danger was far from over.

As the attack continued, another machine gun jammed and stopped firing. Basilone rushed to the weapon and worked quickly to repair it while bullets snapped through the air around him.

Once the gun was working again, he jumped behind it and opened fire.

The Japanese soldiers kept attacking.

But Basilone's guns kept firing.

For hours, the battle raged in the darkness. Basilone moved from position to position, repairing weapons, delivering ammunition, and firing the guns himself when needed.

Again and again, he helped keep the Marine defense alive.

By the time the fighting finally slowed near dawn, the ground in front of the Marine lines was covered with the evidence of the fierce battle.

The Japanese attack had been stopped.

And Sergeant John Basilone had played a crucial role in holding the line.

A Marine Called "Manila John"

When the sun rose over Guadalcanal the next morning, the Marines could finally see the battlefield clearly.

The jungle in front of their defensive lines showed the signs of the fierce fighting from the night before. The Japanese attack had been stopped, but the battle had pushed the Marines to their limits.

Many soldiers had fought bravely during the long night.

But the story of **Sergeant John Basilone** quickly spread through the Marine lines.

Marines talked about how he had kept the machine guns firing when the situation seemed hopeless. They told how he had run through enemy fire to bring ammunition to the gunners and repaired damaged weapons while the battle raged around him.

Without those machine guns, the defensive line might have collapsed.

Among the Marines, Basilone was already known by his nickname—**"Manila John."** The name came from the time he had spent serving in the Philippines before joining the Marine Corps.

After the battle on Guadalcanal, the nickname became even more widely known.

To the Marines who had fought beside him that night, Basilone had shown remarkable courage and determination. He had refused to give up even when the fighting grew desperate and many of his fellow soldiers were wounded.

Because of his actions, the Marine defensive line held.

The Japanese attack had failed.

And the battle for Guadalcanal would continue—but the bravery shown that night would soon become known far beyond the island.

The Medal of Honor

News of the battle on Lunga Area, Guadalcanal eventually reached military leaders far from the island.

Reports described how Sergeant John Basilone had kept the Marine machine guns firing during the intense Japanese attack. His actions had helped stop the enemy advance and protect the defensive line around Henderson Field.

Military officials carefully reviewed what had happened during the battle.

For his courage and leadership, John Basilone was awarded the **Medal of Honor**, the highest military award given by the United States.

Basilone was the first enlisted Marine in WWII to receive the Medal of Honor, the Navy Cross, and the Purple Heart.

In **1943**, Basilone traveled back to the United States to receive the medal and share his story with the American public. Newspapers wrote about the Marine who had helped hold the line during one of the fiercest battles in the Pacific.

Basilone became one of the most well-known American heroes of the war.

Crowds gathered to see him during war bond events, where he helped raise money to support the military effort.

But Basilone never felt comfortable being called a hero.

While he appreciated the recognition, he believed the real heroes were the Marines still fighting overseas.

And before long, he would make a decision that surprised many people.

He asked to return to the battlefield.

Returning to the Fight

After receiving the Medal of Honor, John Basilone spent months traveling across the United States.

During the war, the government encouraged famous soldiers to help sell **war bonds**, which helped raise money to support the military. Large crowds gathered to hear Basilone speak about his experiences on Guadalcanal.

Many people believed Basilone had already done more than enough.

But Basilone felt uneasy staying far from the battlefield while other Marines continued fighting overseas.

Instead of remaining in the United States, he asked for permission to **return to combat**.

In late 1944, Basilone rejoined the Marines preparing for another dangerous mission in the Pacific.

That mission would take him to the island of **Iwo Jima**.

On **February 19, 1945**, American forces began landing on the island's black volcanic beaches. Japanese defenders had built a network of bunkers and gun positions across the island, making the invasion extremely dangerous.

During the fierce fighting on the first day of the battle, Basilone helped lead an attack against heavily defended enemy positions.

While pushing forward with his fellow Marines, **Sergeant John Basilone was killed in action**.

He was only **28 years old**.

John Basilone's courage helped save lives on Guadalcanal and inspired countless Marines who served beside him.

His story reminds us that bravery is not just about a single moment in battle.

Sometimes it is about choosing to stand with others, even when the danger is great.

And Sergeant John Basilone chose that path more than once.

Rodger Young

The Soldier Who Charged Alone

The jungle was thick and quiet.

Tall trees and tangled vines surrounded the American soldiers as they moved carefully through the undergrowth on the island of **New Georgia** in the Solomon Islands.

Visibility was poor. The dense jungle made it difficult to see more than a short distance ahead.

Suddenly, the silence shattered.

A burst of **machine-gun fire** ripped through the trees.

Bullets slammed into the ground and tore through the vegetation as the American soldiers dove for cover. The Japanese machine gun was hidden somewhere up the hill, firing down on the advancing troops.

The soldiers tried to return fire, but the enemy position was well hidden.

Anyone who tried to move forward risked being hit.

Among the soldiers pinned down by the gunfire was **Private Rodger Young**, a soldier in the **148th Infantry Regiment, 37th Infantry Division**.

Young had already been wounded during the fight.

But as the machine gun continued firing, he realized something dangerous.

If the gun was not stopped, the entire platoon could be trapped.

Then Young made a decision.

Despite his wounds, he rose from cover and began moving toward the enemy gun.

A Quiet Boy from Ohio

Rodger Wilton Young was born on **April 28, 1918**, in **Tiffin, Ohio**.

He grew up in the small town of **Clyde**, where life was simple and close-knit. Young was known as a quiet and determined boy who rarely sought attention.

As a child, he faced a serious challenge.

A swimming accident left him with **damaged hearing**, which made it difficult for him to hear clearly for the rest of his life.

Because of this injury, Young was sometimes teased by other children. But those who knew him best remembered something else about him—his persistence.

Rodger Young never liked to quit.

When the United States entered **World War II** after the attack on Pearl Harbor in 1941, millions of young Americans joined the military to serve their country.

Young volunteered for the **United States Army**.

After completing his training, he was assigned to the **37th Infantry Division**, a unit that would soon be sent to fight in the Pacific.

Rodger Young portrait

Private Young was originally a sergeant, but he requested a voluntary demotion to private because he feared his deafness would lead to an error that might get his squad killed.

In 1943, Young and his fellow soldiers arrived in the **Solomon Islands**, where American forces were working to push Japanese troops off a chain of jungle-covered islands.

One of those islands was **New Georgia**.

And it was there that Rodger Young would face the moment that made him a hero.

The Battle for New Georgia

By 1943, the war in the Pacific had become a long and difficult fight across many islands.

After the intense battles on **Guadalcanal**, American forces began pushing north through the **Solomon Islands**. Their goal was to capture key islands that could be used for airfields and naval bases.

Each island brought them closer to Japanese strongholds farther north.

One of the next targets was the island of **New Georgia**.

NEW GEORGIA: 1943

Like many islands in the region, New Georgia was covered in thick jungle, steep hills, and muddy ground. Fighting there was extremely difficult. Soldiers often had to move slowly through dense vegetation where visibility was poor and enemy positions were hard to detect.

The Japanese defenders used the terrain to their advantage.

They built hidden defensive positions in the jungle, including **machine-gun nests and fortified bunkers** that could fire on advancing troops.

American soldiers had to fight their way through these positions one by one.

In **July 1943**, the **37th Infantry Division** began pushing inland across the island. The fighting was fierce as soldiers tried to advance through the jungle toward Japanese defensive lines.

During one of these advances on **July 31, 1943**, Rodger Young's platoon moved forward to adjust their battalion's defensive position for the night.

But as they approached a wooded ridge, the soldiers suddenly came under heavy fire.

A Japanese machine gun hidden higher on the hill had spotted them.

And the weapon was about to stop the entire platoon in its tracks.

Pinned Down

The machine gun fired again.

Bullets tore through the trees and slammed into the ground around the American soldiers. The hidden Japanese gun sat on higher ground, giving its crew a clear view of the jungle below.

Rodger Young's platoon had walked straight into the danger.

The soldiers quickly dropped to the ground, trying to find whatever cover they could among the trees and thick undergrowth. Some fired their rifles toward the hill, hoping to force the enemy gunner to duck down.

But the machine gun kept firing.

Its position was well hidden, and the steep hillside made it extremely dangerous for anyone to move forward.

Then Young was hit.

A burst of enemy fire struck him, wounding him early in the fight. Despite the injury, he refused to leave the battlefield and stayed with his fellow soldiers as the gunfire continued.

The platoon was now trapped.

With the machine gun firing down on them, moving forward seemed impossible. But staying where they were was just as dangerous.

Eventually, the order was given to **withdraw**.

The soldiers began preparing to pull back through the jungle to safer ground.

But Rodger Young realized something important.

If the machine gun kept firing, the platoon might never escape.

There was only one way to stop it.

Someone would have to attack the gun.

And Rodger Young decided that someone would be him.

The Charge

Rodger Young did not wait for someone else to act.

Despite the pain from his wound, he began crawling forward through the jungle toward the machine gun.

The enemy weapon sat about **75 yards away** on higher ground. From that position, it could fire down toward the American soldiers below.

As Young moved forward, the machine gun fired again.

Bullets struck the ground around him, and another burst of fire wounded him a second time. But Young kept going.

Instead of turning back, he rose to his feet and began moving faster.

The enemy gunners spotted him immediately.

The machine gun focused its fire on Young as he advanced. By drawing the gun's attention toward himself, he gave the other soldiers in his platoon a chance to move.

Even though he had already been wounded, Young continued pushing toward the enemy position.

He crept closer and closer through the jungle.

When he finally reached the machine-gun nest, he began throwing **hand grenades** toward the enemy soldiers defending it.

The grenades exploded around the position.

Young was hit again during the attack.

But his actions had done exactly what he hoped.

The machine gun stopped firing.

With the enemy gun disabled and its crew scattered, the pressure on the American soldiers below finally lifted.

Rodger Young's platoon was able to withdraw from the dangerous position.

His courage had saved the lives of many of his fellow soldiers.

But Rodger Young would not survive the battle.

"Don't Give Up"

The fighting around the hillside slowly began to quiet.

With the Japanese machine gun no longer firing, the American soldiers were finally able to pull back through the jungle. The dangerous position that had trapped the platoon was no longer holding them in place.

Rodger Young lay wounded near the enemy position he had attacked.

The soldiers around him understood what he had done. By charging the machine gun and drawing its fire toward himself, Young had given the rest of the platoon a chance to escape.

Before he died, Young spoke a few final words to the soldiers nearby.

"**They got me. But don't give up.**"

Those words stayed with the men who had fought beside him that day.

Rodger Young had known the danger when he moved toward the machine gun. Even after being wounded, he chose to continue forward because he understood what might happen if the gun kept firing.

His courage had made the difference.

Because of his actions, many of his fellow soldiers were able to make it safely away from the battlefield.

In the years that followed, the story of Rodger Young's bravery would become one of the most well-known examples of sacrifice during World War II.

The Medal of Honor

The story of Rodger Young's bravery soon spread through the American military.

Reports from the battlefield described how the wounded soldier had crawled forward under heavy enemy fire to attack the Japanese machine gun that was trapping his platoon. By drawing the gun's attention toward himself and destroying the position, he allowed the rest of the soldiers to escape.

Military leaders carefully reviewed the events of the battle.

For his courage and sacrifice, **Private Rodger Wilton Young** was awarded the **Medal of Honor**, the highest military award given by the United States.

The award was given **posthumously**, meaning it was presented after his death. The official citation described how Young continued advancing toward the enemy machine gun even after being wounded and knowingly placed himself in extreme danger to protect his fellow soldiers.

During World War II, the story of Rodger Young became widely known across the United States.

His bravery inspired soldiers and civilians alike. Songs, newspaper articles, and stories told of the quiet soldier who had charged a machine gun to save others.

Rodger Young had shown a powerful kind of courage—the willingness to risk everything for the people beside him.

Why His Courage Still Matters

Rodger Young never expected to become famous.

He was known by friends and fellow soldiers as a quiet and determined man who simply tried to do his duty. Yet when the moment came during the battle on New Georgia, he chose to act with extraordinary courage.

Young knew the danger when he began moving toward the machine gun. He had already been wounded, and the enemy position was still firing.

But he also understood something else.

If the gun kept firing, the rest of his platoon might not escape.

So he moved forward anyway.

Rodger Young's actions remind us that bravery is not always loud or dramatic. Sometimes it comes from people who quietly decide to help others, even when the risk is great.

Think about the moment when Young stood up and began moving toward the machine gun.

What kind of courage does it take to make a choice like that?

Rodger Young did not act because he wanted recognition or praise. He acted because the soldiers beside him needed help.

And because of that choice, many of them lived to see another day.

His courage is a reminder that even one person can make a difference when others are counting on them.

Henry Erwin

The Airman Who Saved His Crew

The bomber roared high above the clouds.

Inside the massive **B-29 Superfortress**, the crew prepared for their mission over Japan. The plane was part of a group of bombers flying toward their target.

Among the crew was **Staff Sergeant Henry "Red" Erwin**, the plane's **radio operator**.

As the aircraft approached its target near **Koriyama, Japan**, Erwin prepared to perform an important job.

His task was to release **phosphorus smoke bombs** from the aircraft.

These bombs were used to mark targets so that other bombers flying behind them could see exactly where to drop their bombs.

Erwin pulled the release cord.

But something went terribly wrong.

Instead of dropping away from the plane, the smoke bomb malfunctioned.

The burning bomb shot back into the aircraft.

Instantly, the inside of the bomber filled with **blinding white smoke and burning phosphorus**. The fiery chemical sprayed across the metal floor, sending thick smoke through the narrow cabin.

The bomb burned with incredible heat.

If it remained inside the aircraft, the fire could destroy the plane—and everyone aboard.

The crew was suddenly in terrible danger.

And Henry "Red" Erwin was standing only a few feet away from the burning bomb.

A Young Man from Alabama

Henry Eugene Erwin was born on **May 8, 1921**, in **Adamsville, Alabama**.

He grew up in a small community near **Birmingham**, where life was shaped by hard work and close family ties. Like many young people growing up during the years of the **Great Depression**, Erwin learned early that determination and perseverance were important parts of everyday life.

Friends and family often described him as dependable and steady—someone who could be trusted to do the right thing, even when situations became difficult.

When the United States entered **World War II** after the attack on Pearl Harbor in December 1941, millions of Americans stepped forward to serve.

Erwin joined the **U.S. Army Air Forces**, the branch of the military responsible for operating the nation's growing fleet of military aircraft.

Henry Erwin portrait

After completing his training, he became a **radio operator** aboard a **B-29 Superfortress**, one of the largest and most powerful bombers used during the war.

The B-29 carried large crews and could fly long distances across the Pacific Ocean to reach targets in Japan.

Serving on one of these aircraft meant that every crew member had an important responsibility.

And during one mission in **April 1945**, Henry "Red" Erwin would face a moment that required extraordinary courage.

Bombing Missions Over Japan

By 1945, World War II in the Pacific had entered its final and most intense stage.

American forces had fought their way across the Pacific Ocean for nearly four years, capturing islands that allowed them to move closer and closer to **Japan itself.**

One of the most important weapons in this stage of the war was the **B-29 Superfortress** bomber.

The B-29 was one of the most advanced aircraft of its time. It could fly high above enemy defenses and travel long distances across the ocean while carrying large numbers of bombs.

From bases on islands such as **Guam**, American bombers began launching long missions over the Japanese mainland.

These missions often involved large formations of aircraft flying together toward the same target.

But finding the exact target from thousands of feet in the air could be difficult.

To solve this problem, certain aircraft were assigned to drop **phosphorus smoke bombs** that marked the target area. The bright smoke made it easier for the bombers flying behind them to see exactly where to release their bombs.

Henry "Red" Erwin's aircraft was one of those planes.

His job as the **radio operator** also included the dangerous responsibility of releasing those smoke bombs during the mission.

On **April 12, 1945**, as his B-29 approached its target near **Koriyama, Japan**, Erwin prepared to release one of the marking bombs.

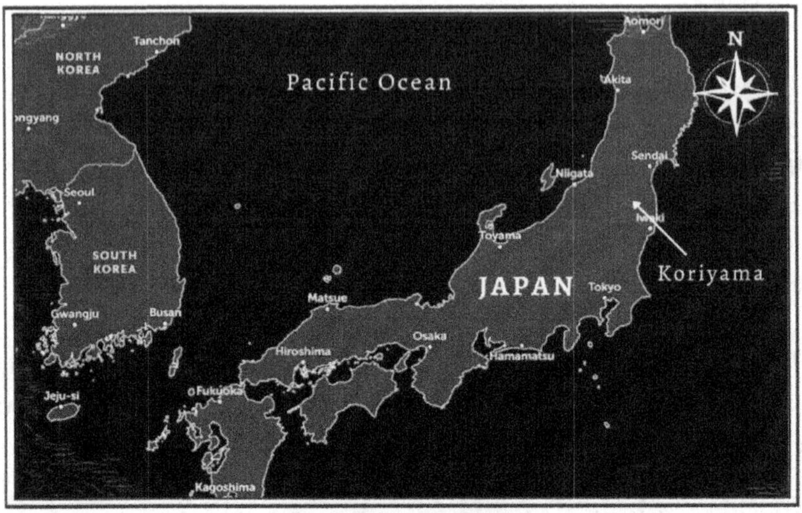

KORIYAMA: 1945

But the moment he pulled the release cord, something went terribly wrong.

The burning bomb did not fall away from the aircraft.

Instead, it came flying back inside the plane.

Fire Inside the Bomber

The burning bomb landed inside the aircraft.

White phosphorus burst across the metal floor, spraying bright sparks and thick smoke through the narrow passageway of the B-29. The chemical burned with incredible heat, producing blinding white light and choking smoke.

Almost instantly, the inside of the bomber filled with fumes.

The crew struggled to see. Smoke spread through the aircraft and began drifting toward the cockpit where the pilots were flying the plane.

If the smoke reached the cockpit, the pilots could lose control of the aircraft.

The bomb itself burned with such intense heat that it could easily start a fire inside the bomber's interior. If that happened, the entire aircraft—and everyone aboard—could be lost.

Henry "Red" Erwin was standing only a few feet away when the bomb exploded inside the plane.

The burning phosphorus struck him directly.

The heat burned his face and body, and the bright chemical flames made it difficult for him to see. Smoke quickly filled the compartment as the bomb continued to burn on the floor.

But Erwin understood the danger immediately.

If the bomb stayed inside the aircraft, the crew would not survive the mission.

There was only one way to save the plane.

Someone had to remove the burning bomb.

And Henry "Red" Erwin decided he would do it himself.

Carrying the Fire

Henry "Red" Erwin knew there was almost no time.

The burning phosphorus bomb continued to spit sparks and thick smoke across the metal floor of the aircraft. The white smoke was spreading quickly toward the cockpit where the pilots were trying to keep the massive bomber steady.

If the smoke filled the cockpit, the pilots could lose sight of their instruments.

The aircraft could crash.

Ignoring the intense heat and smoke, Erwin moved toward the burning bomb.

The white phosphorus burned at temperatures between 1,100- and 1,500-degrees Fahrenheit, requiring swift action to save the crew.

The phosphorus flames were already causing severe burns, but Erwin bent down and **picked up the bomb with his bare hands**.

The heat was overwhelming.

As he lifted the bomb, the burning chemical splashed across his face and body. The smoke made it nearly impossible to see, and the pain was immediate and intense.

But Erwin did not drop the bomb.

Instead, he began moving through the aircraft.

Holding the burning bomb against his body to keep it from falling, he struggled through the narrow passageway of the B-29 toward the cockpit.

The thick smoke filled the plane as he moved forward step by step, all while the plane continued to fall towards the ocean.

Finally, Erwin reached the cockpit.

With the last of his strength, he **threw the burning bomb out the copilot's window**, sending it tumbling away into the sky below.

The deadly smoke immediately began clearing from the aircraft.

The bomber—and everyone aboard it—had been saved. The pilot was able to recover the plane at a mere 300 feet above the ocean, preventing a fatal crash.

But Henry "Red" Erwin had suffered terrible burns during the rescue.

The Crew He Saved

Once the burning bomb was gone, the smoke inside the aircraft began to clear.

The pilots could finally see their instruments again, and the bomber remained under control. The danger that had threatened to destroy the plane had passed.

But the crew quickly realized what Henry "Red" Erwin had endured.

The burning phosphorus had caused **severe burns** across much of his body. The bright chemical flames had injured his face, his hands, and his arms as he carried the bomb through the aircraft.

Despite his terrible injuries, Erwin had refused to drop the bomb until he could throw it safely out of the plane.

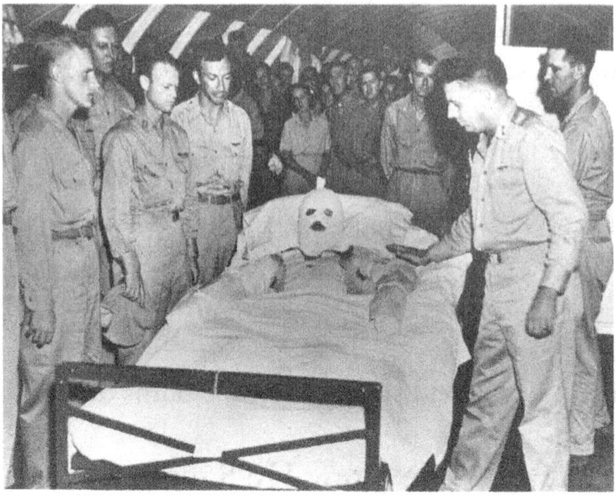

Henry Erwin recovers from his burns. Erwin later recovered and lived a full life until his passing in 2002 at 80 years old.

The crew immediately began helping him as the aircraft turned toward the nearest base.

Doctors were waiting when the bomber landed.

Erwin's injuries were extremely serious. The burns covered much of his body, and the damage to his face and eyes was especially severe. At first, many doctors believed that he might not survive.

But Henry "Red" Erwin had already shown incredible determination.

Just as he had refused to give up inside the burning aircraft, he refused to give up during the long and difficult recovery that followed.

Slowly, he began to heal.

And the crew members whose lives he had saved would never forget what he had done.

The Medal of Honor

The story of Henry "Red" Erwin's bravery quickly spread through the U.S. Army Air Forces.

Reports described how the radio operator had carried a burning phosphorus bomb through the smoke-filled aircraft to save his fellow crew members. Without his actions, the bomber might have been destroyed in midair.

Military leaders reviewed what had happened during the mission.

They concluded that Erwin had shown extraordinary courage by risking his life to protect the rest of the crew.

For his actions, **Staff Sergeant Henry Eugene Erwin** was awarded the **Medal of Honor**, the highest military award given by the United States.

Because his injuries were so severe, Erwin was still recovering in a hospital when the medal was presented to him.

On **April 19, 1945**, Erwin was presented with the Medal of Honor by Maj. Gen. Willis H. Hale, while Gen. Curtis LeMay drove the urgent effort to get the medal there.

Even though he was still badly injured, Erwin was able to receive the nation's highest honor for his courage.

His actions had saved the lives of every member of the bomber's crew.

Why His Courage Still Matters

Henry "Red" Erwin was not a pilot or a commander.

He was a radio operator whose job was to help keep the bomber crew connected and carry out important tasks during their missions. Yet when danger suddenly filled the aircraft, he made a decision that required extraordinary courage.

The burning phosphorus bomb could have destroyed the plane and killed everyone aboard.

Erwin understood that immediately.

Even though the flames severely burned him, he chose to pick up the bomb and carry it through the aircraft to throw it out of the cockpit window.

What kind of courage does it take to make a choice like that?

Erwin could have stepped away from the burning bomb and tried to protect himself. Instead, he acted to protect the people around him.

Because of that choice, the bomber stayed in the air and the entire crew survived the mission.

Henry "Red" Erwin's story reminds us that bravery is not always about fighting an enemy.

Sometimes it means risking everything to save others.

And in one moment high above the clouds, Henry Erwin did exactly that.

Jack H. Lucas
The Teenage Hero of Iwo Jima

The ground of **Iwo Jima** was dark and rough beneath the Marines' boots.

Black volcanic sand covered the island's hills and valleys, and the air smelled of smoke and gunpowder. Explosions echoed across the battlefield as American Marines fought their way forward against determined Japanese defenders.

Hidden enemy positions were scattered across the island.

Machine guns fired from bunkers carved into the rocky ground, and rifle shots cracked across the battlefield as Marines moved carefully from one piece of cover to the next.

Among the Marines advancing that day was **Private First-Class Jack Lucas**.

Lucas was part of a small group of Marines moving through a narrow area of broken ground when enemy fire suddenly erupted nearby.

The Marines dropped low and returned fire, trying to spot where the shots were coming from.

Then something small landed in the dirt beside them.

A **grenade**.

Before anyone could react, a second grenade bounced into the same area.

The small metal objects rolled across the ground and stopped only a few feet from the Marines.

Everyone nearby understood what that meant.

In just a few seconds, the grenades would explode.

There was almost no time to escape.

And in that moment, Jack Lucas made a choice that would save the lives of the men beside him.

A Determined Young Marine

Jacklyn Harold Lucas was born on **February 14, 1928**, in the small town of **Plymouth, North Carolina**.

He grew up during the difficult years of the **Great Depression**, when many American families struggled to make ends meet. Like many boys of his generation, Lucas admired the soldiers and sailors who served their country.

But Jack Lucas was not known for being patient.

Friends and family often described him as energetic, adventurous, and sometimes a little stubborn. Once he decided to do something, it was very hard to change his mind.

When the United States entered **World War II** after the attack on **Pearl Harbor** in December 1941, Lucas was only **13 years old**.

Still, he wanted to serve.

As the war continued and stories of battles across the Pacific filled newspapers and radio broadcasts, Lucas became even more determined to join the fight.

But there was a problem.

He was too young.

Lucas tried more than once to join the military, even running away from school in hopes of enlisting. Eventually, while still underage, he managed to enlist in the **U.S. Marine Corps Reserve**.

Lucas was a "stowaway". He was stationed in Hawaii and was so desperate to join the fight that he jumped on a transport ship and hid until it was at sea.

Jack Lucas portrait

By the time he was **17 years old**, Lucas was serving with **Company C, 1st Battalion, 26th Marines, 5th Marine Division**.

Not long after arriving in the Pacific, he would find himself in one of the most dangerous battles of the entire war.

The battle for **Iwo Jima**.

The Battle for Iwo Jima

By early **1945**, World War II in the Pacific had reached a critical stage.

For years, American forces had been fighting their way across the Pacific Ocean, capturing islands one by one from Japanese control. Each new island allowed American ships and aircraft to move closer to **Japan itself**.

One of the next targets was the island of **Iwo Jima**.

The island was small, only a few miles across, but it held two important **airfields** used by Japanese aircraft. These airfields allowed enemy fighters to attack American bombers flying missions over Japan.

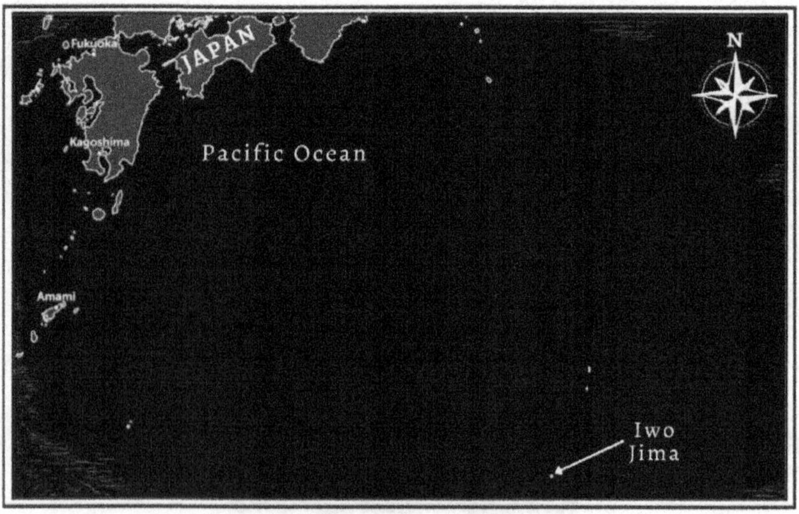

IWO JIMA: 1945

If the United States could capture the island, those airfields could instead be used to protect American bombers and provide emergency landing sites for damaged aircraft returning from combat missions.

But the Japanese defenders had spent months preparing the island for battle.

They built a network of **bunkers, tunnels, and hidden gun positions** across the rocky terrain. Many of these positions were buried deep underground, making them extremely difficult to destroy.

On **February 19, 1945**, thousands of U.S. Marines landed on Iwo Jima's black volcanic beaches.

The fighting that followed was some of the fiercest of the entire war.

Just one day after the invasion began, Jack Lucas and his fellow Marines were already moving forward across the battlefield.

And in one sudden moment, the danger around them would change everything.

A Split-Second Decision

The two grenades rolled to a stop in the dirt.

For a moment, everything seemed to slow down.

Jack Lucas and the Marines around him knew exactly what they were looking at. Japanese grenades were designed to explode only a few seconds after being thrown.

There was almost no time to react.

The Marines were lying close together in the narrow ravine, using the uneven ground for cover as enemy fire continued snapping through the air above them. If the grenades exploded where they had landed, the blast would strike everyone nearby.

The soldiers had only seconds to decide what to do.

Running away was nearly impossible. The grenades were too close, and the rocky ground made it difficult to move quickly.

In that instant, Jack Lucas understood the danger facing the Marines beside him.

If someone did not act, the explosion could kill them all.

Lucas did not hesitate.

He lunged forward.

Throwing himself across the grenades, he pressed his body down on top of them, using himself as a shield to protect the other Marines.

The men around him barely had time to react.

Then the first grenade exploded.

The Explosion

The grenade exploded with a powerful blast. Although the second grenade was a dud (which he did not know at the time), the first had done its fair share of damage.

The force of the explosions shook the ground and sent fragments of metal flying in every direction. Dirt and smoke burst into the air as the Marines nearby were thrown backward by the shock of the blast.

But the explosion had been mostly absorbed by the body of **Jack Lucas**.

The young Marine lay badly wounded on the ground.

The grenade had filled his body with pieces of metal called **shrapnel**. Fragments tore into his chest, arms, legs, and face. Later, doctors would discover that more than **two hundred pieces of shrapnel** had struck him.

The Marines beside him were stunned.

Only moments earlier, the grenades had landed just feet away from them. Without Lucas's quick action, the explosion could have killed or severely wounded everyone nearby.

Instead, they were still alive.

The soldiers rushed to help him.

Medics quickly arrived and began treating Lucas's terrible wounds while the battle continued around them. Despite the damage from the explosion, the young Marine was still breathing.

He had survived the blast that he had thrown himself onto.

Because of his decision in that single moment, the Marines around him had survived as well.

And the story of what Jack Lucas had done would soon spread across the battlefield.

Fighting to Recover

Jack Lucas's injuries were severe.

After the explosion, medics rushed him away from the battlefield as quickly as they could. The young Marine had been struck by hundreds of small pieces of shrapnel from the grenades.

Doctors worked carefully to remove as many of the metal fragments as possible. The injuries covered much of his body, and the recovery would take a long time.

At first, many people were not sure if Lucas would survive.

But the same determination that had pushed him to join the Marines while still underage now helped him through the long months of healing.

Slowly, he began to recover.

It was a difficult process. Doctors performed many surgeries to remove pieces of shrapnel, and Lucas spent months in hospitals while his body healed.

The Marines who had been beside him on Iwo Jima never forgot what had happened that day.

They knew exactly why they were still alive.

In one brief moment during the battle, Jack Lucas had chosen to place himself in danger so that the others around him could survive.

His courage had saved their lives.

And soon, the nation would hear the story of what the young Marine had done on the battlefield.

The Medal of Honor

The story of Jack Lucas's bravery soon spread beyond the battlefield of Iwo Jima.

Military leaders reviewed the events that had taken place during the fierce fighting on the island. Reports from the Marines who had been with Lucas described how he had thrown himself onto the two grenades to protect the soldiers beside him.

They knew exactly what would have happened if he had not acted.

For his extraordinary courage, **Private First-Class Jacklyn Harold Lucas** was awarded the **Medal of Honor**, the highest military award given by the United States.

Lucas was still recovering from his injuries when he received the honor.

On **October 5, 1945, President Harry S. Truman** presented the Medal of Honor to Lucas during a ceremony at the **White House**.

Jack Lucas receives the Medal of Honor

At just **17 years old**, Jack Lucas became the one of the youngest Marines to receive the Medal of Honor during World War II.

The teenager who had once been too young to serve in the military was now recognized as one of the bravest soldiers of the war.

But Lucas never thought of himself as a hero.

He believed he had simply done what was necessary to protect the Marines beside him.

Why His Courage Still Matters

Jack Lucas was only **17 years old** when he faced one of the most dangerous moments of the war.

Many teenagers his age were still in school or working their first jobs. But on the battlefield of Iwo Jima, Lucas found himself in a situation where a decision had to be made in only a few seconds.

Two grenades had landed beside the Marines around him.

Lucas understood what that meant.

If the grenades exploded where they had fallen, the blast could kill the men beside him.

So he acted.

Without hesitation, he used his own body to shield the others from the explosion.

What kind of courage does it take for someone so young to make a decision like that?

Jack Lucas's story reminds us that bravery is not determined by age or experience.

Sometimes courage appears in a single moment when someone chooses to protect others, even when the risk is great.

And on that battlefield in 1945, Jack Lucas made that choice.

Because of him, the Marines beside him lived to see another day.

Jack Lummus

The Football Star Who Led the Charge

The Marines were pushing forward across the rough ground of **Iwo Jima**.

Smoke drifted across the battlefield as explosions echoed through the rocky hills. Hidden Japanese defenses were scattered across the island, and every step forward was dangerous.

Among the Marines leading the advance was **First Lieutenant Jack Lummus**.

Lummus commanded a platoon of Marines moving toward a series of enemy positions dug into the rocky ground ahead of them.

The Japanese defenders had prepared carefully.

Machine guns fired from hidden bunkers while rifles cracked across the battlefield. The Marines moved forward carefully, using craters and rocks for cover as they tried to push the defenders back.

Suddenly, an explosion ripped through the air.

A hidden **land mine** buried beneath the ground had been triggered.

The blast knocked Lummus to the ground.

For a moment, everything seemed to stop.

The Marines nearby rushed toward their wounded leader. The explosion had caused terrible injuries to his legs, and the pain must have been overwhelming.

But even as the battle continued around him, Jack Lummus refused to give up.

Instead of ordering his men to fall back, he pushed himself up and urged them to keep advancing.

The fight for the island was not over.

And neither was Jack Lummus.

From Football Field to Battlefield

Jack Lummus was born on **October 22, 1915**, in **Ennis, Texas**. He was the only son in a family of four children.

As a young man, Lummus loved sports—especially **football**. He grew up strong, competitive, and determined, qualities that made him stand out on the playing field.

After high school, Lummus became a talented athlete and eventually played **professional football in the National Football League (NFL)**. He played as a **running back** for the **New York Giants**, one of the top teams in professional football at the time.

But when the United States entered **World War II** after the attack on **Pearl Harbor** in 1941, many athletes and young men across the country felt a responsibility to serve.

Jack Lummus was one of them.

He left his football career behind and joined the **United States Marine Corps Reserve**.

During his training, Lummus quickly proved himself to be a strong leader. His determination and discipline helped him rise through the ranks, and he was eventually commissioned as a **first lieutenant**.

Jack Lummus portrait

By early **1945**, Lummus was leading a platoon of Marines in **Company E, 2nd Battalion, 27th Marines, 5th Marine Division**.

Not long after arriving in the Pacific, his unit would be sent into one of the toughest battles of the war.

The battle for **Iwo Jima**.

Fighting Across the Island

By **March 1945**, the battle for **Iwo Jima** had already been raging for more than two weeks.

American Marines had landed on the island's black volcanic beaches on **February 19**, but capturing the island was proving far more difficult than anyone had expected.

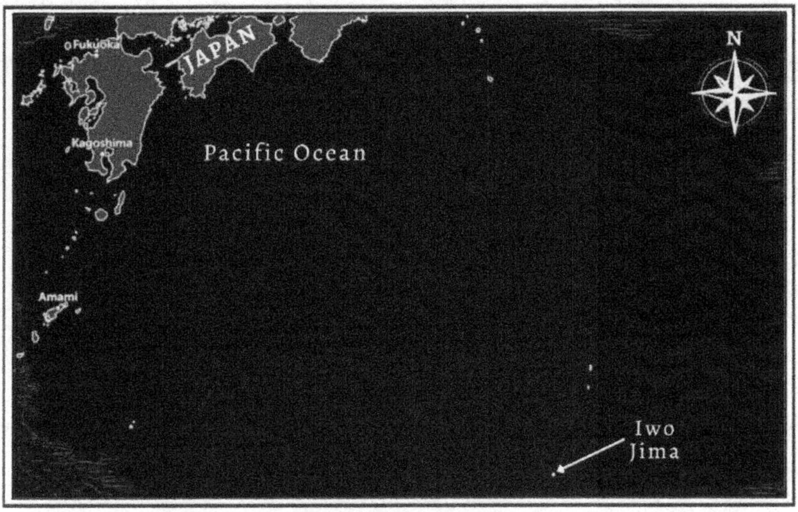

IWO JIMA: 1945

Japanese defenders had built an enormous network of **bunkers, tunnels, and hidden firing positions** across the island. Many of these positions were buried deep underground and connected by tunnels that allowed soldiers to move unseen from one location to another.

Instead of defending only the beaches, Japanese forces waited until the Marines moved inland.

Then the fighting became even more intense.

Every hill, ridge, and rocky valley could hide machine guns, snipers, or artillery positions. Marines often had to attack these defenses one at a time, moving forward slowly while facing constant enemy fire.

By early March, the battle had turned into a brutal struggle for control of the island's northern hills.

It was in this dangerous landscape that **First Lieutenant Jack Lummus** and his platoon were ordered to advance against a series of heavily defended Japanese positions.

The enemy defenses were strong.

But the Marines had to keep moving forward.

And during one of those advances, Jack Lummus would show extraordinary courage while leading his men into battle.

Leading the Attack

The Marines moved carefully across the rocky ground of northern **Iwo Jima**.

First Lieutenant **Jack Lummus** led his platoon forward as they approached a series of Japanese defensive positions built into the hills ahead. Hidden bunkers and gun emplacements protected the defenders, and every advance brought the Marines under heavy fire.

Machine guns rattled across the battlefield.

Bullets struck the ground and rocks around the Marines as they pushed forward. Lummus moved among his men, encouraging them to keep advancing while directing their fire toward the enemy positions.

The fighting was intense.

Several times the platoon was forced to take cover as grenades exploded and rifle fire snapped through the air. But each time, Lummus pushed his Marines forward again.

They were determined to break through the Japanese defenses.

As the platoon advanced, Lummus spotted one of the enemy positions ahead and began moving his men into position to attack it.

That was when disaster struck.

Hidden beneath the rocky ground was a **Japanese land mine**.

As Lummus stepped forward to continue the attack, the mine suddenly exploded beneath him.

The blast threw him to the ground and severely injured his legs.

The Marines nearby rushed toward their wounded leader.

But even as the battle continued around him, Jack Lummus refused to give up.

Refusing to Stop

The explosion had caused terrible injuries to **Jack Lummus's legs**.

The Marines who rushed to help him could see how serious the wounds were. The blast from the land mine had severely damaged his lower legs, and the pain must have been overwhelming.

But even as the battle raged around them, Lummus refused to leave the fight.

Instead of allowing his Marines to stop and focus on him, he urged them to keep moving forward.

The enemy positions ahead were still firing, and the attack could not stop.

Despite his injuries, Lummus continued encouraging his men to press the attack against the Japanese defenses. His determination helped keep the Marines focused as they fought through the enemy positions.

Inspired by their leader's courage, the Marines pushed forward.

They attacked the remaining bunkers and gun positions, gradually breaking through the Japanese defenses.

Throughout the fight, Lummus continued directing his men and urging them onward even though he had been gravely wounded.

His leadership helped the platoon continue the attack during one of the most dangerous moments of the battle.

But the injuries from the explosion were too severe.

Jack Lummus had given everything to help his Marines succeed in their mission.

A Leader to the End

The Marines around Jack Lummus could see how serious his injuries were.

Medics quickly moved toward him as the battle continued across the rocky hills of Iwo Jima. The explosion from the land mine had caused devastating wounds to his legs, and the situation was extremely serious.

Even so, Lummus continued thinking about his men.

Instead of focusing on his own injuries, he urged the Marines around him to keep moving forward and complete their mission. His determination helped inspire the soldiers who were still fighting nearby.

The attack continued.

The Marines pushed forward against the remaining Japanese positions, determined to finish the job that their leader had started.

But Jack Lummus's injuries were too severe.

Despite the efforts of the medics who tried to help him, the young Marine officer died from his wounds later that day on **March 8, 1945**.

Shortly before he died from fatal land mine wounds, he told the medic, "Well, Doc, it looks like the Giants have lost a good end".

The Marines who had fought beside him would never forget what they had seen.

Even after suffering terrible injuries, Jack Lummus had continued to lead and encourage his men.

He had shown the kind of courage and leadership that soldiers remember for the rest of their lives.

The Medal of Honor

After the battle on **Iwo Jima**, reports describing Jack Lummus's leadership and bravery were carefully reviewed by military officials.

The Marines who had fought beside him explained how Lummus had continued leading the attack even after the land mine explosion had caused terrible injuries. Instead of allowing the advance to stop, he encouraged his men to continue moving forward and defeat the enemy defenses.

His courage and determination helped his platoon complete their mission during one of the most dangerous moments of the battle.

For his actions, **First Lieutenant Jack Lummus** was awarded the **Medal of Honor**, the highest military award given by the United States.

Because Lummus had died from his injuries, the medal was awarded **posthumously**, meaning it was given after his death.

The official citation recognized how he had continued directing and encouraging his Marines despite being gravely wounded.

Jack Lummus had already shown determination and discipline as an athlete before the war. But on the battlefield of Iwo Jima, he showed an even greater kind of strength—the courage to lead others forward even when facing incredible danger.

His actions remain one of the many examples of extraordinary bravery shown during the battle for the island.

Why His Courage Still Matters

Jack Lummus had already achieved something many young athletes dream about.

He had played **professional football in the NFL**, competing at the highest level of the sport. But when his country needed soldiers during World War II, he chose to leave his football career behind and serve in the **United States Marine Corps**.

On the battlefield of **Iwo Jima**, Lummus showed a different kind of strength.

As a platoon leader, he was responsible for guiding and protecting the Marines under his command. Even when he was badly injured by the explosion of a hidden land mine, he refused to stop leading.

Instead of focusing on his own pain, he encouraged his men to continue the attack and complete their mission.

What kind of courage does it take to keep thinking about others in a moment like that?

Jack Lummus's story reminds us that true leadership means putting the needs of others first.

On that battlefield in 1945, he showed his Marines what it meant to lead with courage.

And because of his example, his story is still remembered today.

Hershel "Woody" Williams

The Flamethrower Assault at Iwo Jima

The Marines were pinned down.

Across the rocky hills of **Iwo Jima**, Japanese defenders fired from hidden bunkers built deep into the volcanic ground. The enemy positions were made of reinforced concrete and stone, carefully designed to survive artillery fire and protect the soldiers inside.

Machine guns rattled from the openings of the bunkers.

Bullets slammed into the rocks and dirt around the Marines, forcing them to stay low behind the little cover they could find.

Every attempt to move forward brought heavy enemy fire.

The bunkers were part of a strong defensive line protecting the Japanese positions ahead. As long as the machine guns kept firing, the Marines would not be able to advance.

Someone would have to destroy the bunkers.

But reaching them meant crossing open ground while under enemy fire.

Among the Marines on that battlefield was **Corporal Hershel "Woody" Williams**.

Williams carried one of the most dangerous weapons used during the war—a **flamethrower**, a heavy device that could send a stream of burning fuel into enemy positions.

The weapon was powerful.

But using it meant getting dangerously close to the bunkers.

As the fighting continued around him, Williams realized what had to be done.

Then he moved forward alone.

A Determined Marine

Hershel Woodrow Williams was born on **October 2, 1923**, in **Quiet Dell, West Virginia**.

He grew up in a rural farming community during the difficult years of the **Great Depression**. Like many boys of that time, Williams spent much of his childhood helping his family and working outdoors.

But when the United States entered **World War II** after the attack on **Pearl Harbor** in 1941, Williams wanted to serve in the military.

At first, however, the Marine Corps would not accept him.

Williams was smaller than many other recruits, and the Marines had strict physical requirements. Determined to serve his country, he worked hard to meet those standards and eventually succeeded in joining the **United States Marine Corps Reserve** in 1943.

During training, Williams learned to operate several different weapons. One of them was the **flamethrower**, a weapon designed to destroy enemy bunkers and fortified positions.

Hershel Williams wearing his Medal of Honor

The flamethrower was extremely dangerous to use.

The Marine carrying it had to move very close to the enemy while under fire. But in the rugged terrain of the Pacific islands, it was sometimes the only weapon powerful enough to destroy hidden defensive positions.

By early **1945**, Williams was serving with **Headquarters Company, 1st Battalion, 21st Marines, 3rd Marine Division**.

And during the brutal battle for **Iwo Jima**, that dangerous weapon would become the key to breaking through the enemy defenses.

Breaking the Defenses

By **February 1945**, the battle for **Iwo Jima** had become one of the toughest fights of World War II.

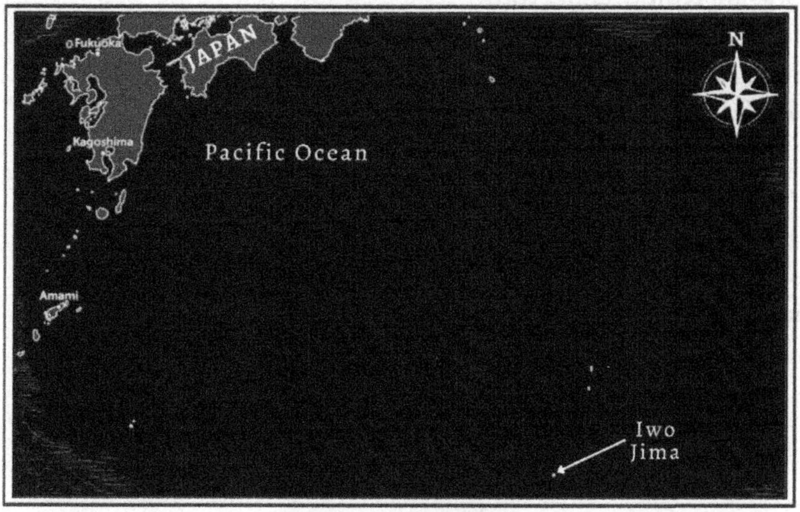

IWO JIMA: 1945

The island was small, but Japanese forces had spent months preparing its defenses. Instead of building large forts above ground, they created a massive network of **bunkers, tunnels, and reinforced pillboxes** hidden within the rocky hills.

Many of these positions were extremely difficult to destroy.

Artillery shells and bombs often bounced off the thick concrete walls, and the defenders could fire their weapons through narrow openings while remaining protected inside.

As Marines tried to move forward across the island, these hidden bunkers caused deadly problems.

Machine guns fired from one position while another bunker nearby provided support. Even if Marines managed to silence one position, another hidden defense might open fire from a different direction.

The only reliable way to destroy these strongpoints was to attack them up close.

This was incredibly dangerous.

Marines sometimes used explosives, grenades, or special weapons designed to burn enemy positions out of their defenses.

One of those weapons was the **flamethrower**.

And during the battle for Iwo Jima, **Corporal Hershel "Woody" Williams** would use that weapon in a way that few soldiers had ever attempted before.

Moving Toward the Bunkers

The enemy bunkers continued firing.

Machine guns rattled across the battlefield, pinning the Marines down behind rocks and shallow craters. Every time the soldiers tried to move forward, the hidden defenders opened fire again.

As long as the bunkers remained active, the Marines would not be able to advance.

Someone had to get close enough to destroy them.

That job fell to **Corporal Hershel "Woody" Williams**.

Carrying his **flamethrower**, Williams began moving forward across the rocky ground. The weapon was heavy, and its fuel tanks made him an easy target for enemy fire.

Bullets struck the ground around him as he crawled toward the first bunker.

Two Marines moved nearby, helping provide cover as Williams worked his way closer to the enemy position.

Reaching the bunker required careful movement. The opening where the machine gun fired was small, and the defenders inside were watching for anyone who tried to approach.

But Williams kept moving.

When he finally reached the bunker, he raised the flamethrower and aimed it toward the opening.

Then he fired.

A burst of burning fuel shot into the enemy position, forcing the defenders inside to abandon their bunker.

But there were many more bunkers ahead.

And Woody Williams was just getting started.

Bunker After Bunker

Destroying the first bunker did not end the fight.

More enemy positions were scattered across the rocky hillside, and machine guns continued firing at the Marines below.

Without waiting for orders, **Woody Williams kept moving forward**.

Carrying his flamethrower, he crawled and ran from cover to cover while enemy bullets struck the ground around him. Each bunker he approached was dangerous. Japanese soldiers inside the defenses knew someone was attacking their positions, and they fired at anyone who came near.

Still, Williams continued his assault.

When he reached the next bunker, he aimed the flamethrower into the opening and fired another burst of burning fuel. Flames rushed into the position, forcing the defenders inside to abandon it.

Then he moved to the next one.

For **hours**, Williams repeated the same dangerous task.

Sometimes other Marines helped cover him with rifle fire. At other moments, Williams attacked the bunkers almost completely alone, moving close enough to use the flamethrower against the openings of the enemy defenses.

Again and again, the bunkers fell silent.

One position after another was destroyed, allowing the Marines to continue advancing across the battlefield.

By the end of the assault, **Woody Williams had helped eliminate several of the most dangerous enemy positions** blocking the Marines' advance.

And his courage had helped turn the battle in their favor.

Breaking Through the Line

By the time the fighting slowed, the hillside looked very different.

Several of the Japanese bunkers that had pinned the Marines down were now silent. The machine guns that had once fired across the battlefield had been destroyed or abandoned.

Because of **Woody Williams's assault**, the Marines were finally able to move forward.

The dangerous defensive line that had stopped their advance had been broken.

Williams had spent **hours moving across open ground**, attacking one bunker after another while exposed to enemy fire. It was exhausting and incredibly dangerous work, but he refused to stop until the enemy defenses were destroyed.

The Marines who had been pinned down earlier could now push ahead and continue the attack across the island.

Williams himself survived the battle.

Like many Marines who fought on Iwo Jima, he faced many difficult days of combat before the island was finally secured. But his actions during that assault had played an important role in helping his fellow Marines move forward during one of the most dangerous moments of the battle.

The courage he showed that day would soon be recognized by the entire nation.

The Medal of Honor

After the battle for **Iwo Jima**, reports describing Hershel "Woody" Williams's actions were carefully reviewed by military leaders.

The Marines who had fought beside him explained how Williams had attacked one enemy bunker after another with his flamethrower while exposed to constant enemy fire. His actions had helped destroy several of the strongest defensive positions blocking the Marines' advance.

Without that assault, the attack might have stalled under the heavy machine-gun fire from the bunkers.

Military officials concluded that Williams had shown extraordinary courage.

For his actions, **Corporal Hershel Woodrow Williams** was awarded the **Medal of Honor**, the highest military award given by the United States.

On **October 5, 1945, President Harry S. Truman** presented the medal to Williams during a ceremony at the **White House**.

The young Marine who had crossed open ground again and again to destroy the enemy bunkers was now recognized as one of the nation's heroes.

Many years later, Williams would become the **last surviving Medal of Honor recipient from the Battle of Iwo Jima**, helping keep the memory of the battle and the Marines who fought there alive for future generations.

His courage on that battlefield remains one of the most remarkable stories from the war.

Why His Courage Still Matters

Hershel "Woody" Williams faced one of the most dangerous jobs on the battlefield.

Destroying enemy bunkers required getting close to heavily defended positions while under fire. The Marines inside those bunkers were trained to protect their defenses and stop anyone who tried to approach.

Yet Williams kept moving forward.

Again and again, he crossed open ground carrying his flamethrower and attacked the enemy positions that were stopping the Marines' advance.

What kind of courage does it take to keep moving forward when the danger is so great?

Williams's actions remind us that bravery is not always about a single moment.

Sometimes it means continuing to do the hard and dangerous work that others depend on, even when the risks are high.

On that battlefield at **Iwo Jima**, Woody Williams refused to stop until the enemy defenses were destroyed and his fellow Marines could move forward.

Because of his determination and courage, the attack succeeded.

And his story continues to remind us how one person's actions can help change the course of a battle.

Richard Bong

America's Greatest Fighter Ace

High above the islands of the **Southwest Pacific**, two fighter planes raced across the sky.

The engines roared as the aircraft twisted and turned through the clouds. One plane dove sharply while the other followed close behind, both pilots trying to gain the advantage.

This was a **dogfight**—a fast and dangerous battle between fighter planes.

In the cockpit of one of the American aircraft sat **Captain Richard Bong** of the United States Army Air Forces.

Bong flew a powerful **P-38 Lightning**, a twin-engine fighter known for its speed and heavy firepower. From his seat inside the narrow cockpit, he carefully watched the enemy aircraft ahead of him.

The Japanese fighter pilot tried to escape.

The plane rolled sharply to one side and then climbed into the sky, hoping to break away from the American pilot behind him.

But Richard Bong stayed close.

He had spent countless hours learning how to fly and fight in the air, and he understood exactly how to position his aircraft for a shot.

Slowly, he moved his fighter into the perfect position behind the enemy plane.

Then Bong pressed the trigger.

Machine guns erupted from the nose of the P-38 as the battle in the sky reached its final moment.

A Boy Who Loved to Fly

Richard Ira Bong was born on **September 24, 1920**, in **Superior, Wisconsin**.

He grew up in a small town near the forests and lakes of northern Wisconsin. As a boy, Bong enjoyed outdoor activities and sports, but he developed a special interest in something that was becoming increasingly exciting during the early 1900s—**airplanes**.

Flying was still a relatively new technology, and many young people were fascinated by the idea of soaring through the sky.

Richard Bong was one of them.

As he grew older, Bong spent more time learning about aircraft and dreaming about becoming a pilot. After finishing high school, he attended **Superior State Teachers College**, but his real goal was to learn how to fly.

In **1941**, Bong joined the **United States Army Air Corps**, the branch of the military responsible for aircraft before the creation of the U.S. Air Force.

Richard Bong poses for a picture from his cockpit

During flight training, Bong quickly showed that he had an unusual level of skill in the air. His instructors noticed that he had excellent control of his aircraft and a strong sense of how to maneuver during aerial combat.

By the time the United States entered **World War II**, Richard Bong was ready to put those skills to the test in one of the most dangerous environments a pilot could face.

The skies above the Pacific battlefield.

Fighting in the Skies

During World War II, control of the skies was just as important as control of land or sea.

Fighter pilots protected bombers, attacked enemy aircraft, and helped support soldiers fighting on the ground. In the Pacific theater, American pilots often flew long missions over jungles, islands, and open ocean.

The enemy they faced was skilled and determined.

Japanese fighter pilots flew fast and agile aircraft, and many had years of combat experience when the war began. Battles between fighter planes could happen suddenly and last only a few minutes as the pilots twisted and turned through the sky trying to gain an advantage.

Pilots who successfully shot down several enemy aircraft were given a special title.

They were called **fighter aces**.

To earn that title, a pilot had to shoot down **five enemy airplanes** in combat.

For most pilots, reaching that number was extremely difficult.

But Richard Bong quickly proved that he was no ordinary pilot.

Flying his **P-38 Lightning** fighter plane in the skies over the Pacific, Bong began to achieve victory after victory against enemy aircraft.

With each successful mission, his reputation as a skilled and fearless pilot continued to grow.

And before long, he would become the most successful American fighter pilot of the entire war.

Victory in the Air

Combat missions over the Pacific were dangerous and unpredictable.

American pilots often flew long distances over jungles, islands, and open ocean, never knowing when enemy aircraft might appear. When Japanese fighters were spotted, the sky could suddenly turn into a fast-moving battle.

Richard Bong quickly proved that he was one of the most skilled pilots in these fights.

Flying his **P-38 Lightning**, Bong used the aircraft's speed and firepower to his advantage. He carefully studied the movement of enemy planes and waited for the right moment to strike.

Then he attacked.

In one battle after another, Bong managed to maneuver his aircraft behind enemy fighters and fire his machine guns at close range. Each successful attack added another enemy aircraft to his record.

By **1943**, Bong had already shot down several enemy planes.

His success made him well known among the other pilots in his squadron. They respected his calm flying style and his ability to stay focused even in the middle of intense combat.

But Bong was far from finished.

With each new mission, he continued adding to his total—slowly building a record that no other American fighter pilot had ever reached before.

Becoming America's Top Ace

As the war in the Pacific continued, Richard Bong kept flying mission after mission.

Each time he entered combat, he faced the same dangers—enemy fighters, anti-aircraft fire, and long flights over remote islands and open ocean. Many pilots completed their tours with only a few victories.

But Bong's record continued to grow.

By **1944**, he had already shot down enough enemy aircraft to become a **fighter ace**, a title given to pilots who destroyed at least five enemy planes in combat.

Bong did much more than that.

Again and again, he proved that he could outmaneuver enemy pilots in the sky. Flying his **P-38 Lightning**, he carefully positioned his aircraft behind enemy fighters and attacked at the right moment.

One victory followed another.

His fellow pilots began to realize that Bong was accomplishing something extraordinary. With each successful mission, he moved closer to breaking the record for the most enemy aircraft shot down by an American pilot.

Eventually, Bong reached **40 aerial victories**.

No other American fighter pilot during World War II would reach that number.

Richard Bong had become **America's top fighter ace**, a pilot whose skill and courage in the air made him one of the most successful combat aviators of the war.

A Hero at Home

Richard Bong's success in the skies made him famous across the United States.

Newspapers wrote about the young fighter pilot who had shot down more enemy aircraft than any other American pilot. His victories became a source of pride and encouragement for people back home who were following the progress of the war.

But Bong's growing record also worried military leaders.

Experienced pilots were extremely valuable, and losing the nation's top ace in combat would have been a serious loss. After reaching his record number of victories, Bong was eventually sent back to the United States for a time.

There, he helped promote **war bonds**, which helped raise money to support the war effort. He also shared his experiences with others and encouraged Americans to continue supporting the soldiers fighting overseas.

Even though he was no longer flying daily combat missions, Bong remained connected to aviation.

Flying had always been his passion.

And his remarkable record in the skies of the Pacific had already secured his place in history as one of the most skilled fighter pilots of World War II.

The Medal of Honor

Richard Bong's achievements in the skies of the Pacific drew the attention of military leaders across the United States.

Reports from his squadron described how the young pilot repeatedly faced enemy aircraft in dangerous aerial battles and helped protect American bombers and troops during critical missions.

His courage, skill, and determination made him one of the most respected fighter pilots of the war.

Because of his extraordinary combat record, **Major Richard Ira Bong** was awarded the **Medal of Honor**, the highest military award given by the United States.

Richard Bong portrait

On **December 12, 1944**, **Gen. Douglas MacArthur** presented the medal to Bong during a ceremony at the **Tacloban air strip in the Philippine Islands**.

Bong's Medal of Honor was officially for his repeated combat missions and eight victories during the specific **October 10–November 15, 1944** period over Borneo and Leyte, not just as a blanket award for all 40 victories.

After his combat service, Bong continued flying as a test pilot while new aircraft were being developed near the end of the war.

Tragically, on **August 6, 1945**, Bong was killed when the experimental jet aircraft he was flying crashed during a test flight in California. He was only **24 years old**.

This would be the same day that the first atomic bomb was dropped on Hiroshima, Japan, marking the beginning of the end of the war.

Even though his life ended far too soon, Richard Bong had already secured his place in history as one of the most remarkable fighter pilots of World War II.

Why His Courage Still Matters

Richard Bong's story is different from many other heroes of World War II.

Instead of fighting on the ground, Bong fought high above the battlefield in fast-moving aerial battles. Each mission required incredible focus, skill, and quick decision-making.

In the air, even a small mistake could be dangerous.

Bong became one of the most successful fighter pilots of the war because he trained hard and stayed calm during combat. He studied how enemy aircraft moved and learned how to position his plane for the right moment to attack.

But his success was not only about flying skill.

Bong understood that his job helped protect others. By defeating enemy fighters, he helped keep American bombers and soldiers safer during their missions.

What kind of dedication does it take to face danger in the sky again and again?

Richard Bong's story reminds us that courage can come from careful preparation, determination, and a willingness to do a difficult job well.

Because of his service and skill, he is remembered as one of the greatest fighter pilots in American history.

Daniel Inouye

The Soldier Who Kept Fighting

The fighting on the hillside had already been fierce.

American soldiers moved carefully up the rugged slope near **San Terenzo, Italy**, while enemy machine guns fired from hidden bunkers above them. Each time the soldiers tried to advance, bursts of gunfire forced them back behind rocks and small patches of cover.

One bunker in particular was causing serious trouble.

The enemy machine gun inside the position had already stopped several attempts to move forward. Anyone who tried to approach the bunker faced a deadly stream of bullets.

Among the soldiers pinned down on the hillside was **Second Lieutenant Daniel Inouye**.

Inouye led a platoon of soldiers from the **442nd Regimental Combat Team**, a famous unit made up mostly of Japanese American soldiers.

Seeing that the machine gun had to be stopped, Inouye began moving toward the bunker.

Bullets struck the ground around him as he crawled closer to the enemy position. When he reached a spot just below the bunker opening, he pulled a **grenade** from his gear.

With a quick motion, Inouye threw the grenade into the enemy position.

The explosion silenced the machine gun.

But the battle on the hillside was far from over.

A Soldier from Hawaii

Daniel Ken Inouye was born on **September 7, 1924**, in **Honolulu, Hawaii**.

As a young boy, Inouye grew up in a close-knit community where many families had roots in Japan. He enjoyed school and sports, and like many boys his age, he dreamed about the future.

But everything changed on **December 7, 1941**.

That morning, Japanese aircraft attacked the American naval base at **Pearl Harbor**, only a short distance from where Inouye lived. The attack brought the United States into World War II.

After the attack, many Japanese Americans faced suspicion and discrimination, even though they were loyal citizens of the United States.

Daniel Inouye wanted to prove his loyalty by serving his country.

When the U.S. Army began forming a new unit made up mostly of Japanese-American soldiers, Inouye volunteered.

He joined the **442nd Regimental Combat Team**, a unit that would eventually become one of the most highly decorated military units in American history.

Daniel Inouye

The soldiers of the 442nd adopted a powerful motto:

"Go for Broke."

It meant giving everything they had and holding nothing back.

Daniel Inouye would soon live out that motto on the battlefield.

The Battle in Italy

By **1945**, the war in Europe was entering its final months.

American and Allied forces were pushing north through **Italy**, fighting to drive German troops from the country. The enemy often defended strong positions in hills and mountains, forcing attacking soldiers to move uphill while under fire.

These battles were extremely difficult.

German defenders built machine-gun nests and bunkers that overlooked the roads and valleys below. From these positions, they could fire down on advancing soldiers and stop attacks before they reached the top of the hills.

During one battle near **San Terenzo, Italy**, soldiers from the **442nd Regimental Combat Team** were ordered to capture a heavily defended ridge.

SAN TERENZO: 1945

But the enemy had prepared their defenses carefully.

Several machine guns were hidden in fortified positions along the hillside. Whenever the American soldiers tried to move forward, the machine guns opened fire, forcing them to take cover.

The attack slowed as the soldiers searched for a way to break through the defenses.

Among the officers leading the assault was **Second Lieutenant Daniel Inouye**.

Seeing that the enemy machine guns had to be destroyed before the attack could continue, Inouye began moving forward toward the enemy positions.

The First Bunker

Daniel Inouye knew the attack could not continue until the enemy machine guns were stopped.

The German soldiers had positioned their weapons carefully along the ridge, allowing them to fire down on anyone who tried to move uphill.

If the machine guns kept firing, the soldiers of the **442nd Regimental Combat Team** would remain pinned down.

So Inouye moved forward.

Keeping low to the ground, he crawled toward the first enemy position while bullets struck the dirt and rocks around him. The machine gun inside the bunker continued firing at the soldiers behind him.

When Inouye reached a spot close enough to attack, he pulled a **grenade** from his gear.

With a quick motion, he threw the grenade into the bunker opening.

The explosion forced the enemy soldiers inside the position to stop firing. Inouye immediately rushed forward and destroyed the bunker, clearing the way for his platoon to move closer to the ridge.

But as he looked up the hillside, he saw another enemy position still firing.

A second machine gun nest was waiting higher on the ridge.

Without hesitation, Inouye began moving toward that position as well.

The battle was far from finished.

The Fight Continues

Daniel Inouye crawled toward the second enemy position.

The German soldiers inside the bunker could see movement on the hillside and opened fire again. Bullets struck the ground and rocks around him as he slowly moved closer to the machine gun nest.

Inouye reached a spot near the bunker and prepared another grenade.

He pulled the pin and threw it into the enemy position.

The explosion destroyed the machine gun nest, clearing another obstacle for the advancing soldiers of the **442nd Regimental Combat Team**.

But as Inouye stood up to continue the attack, a **sniper's bullet struck him in the chest**.

The impact knocked him backward onto the hillside.

Despite the wound, Inouye refused to stop fighting.

He struggled to his feet and continued moving forward toward another enemy bunker that was still firing at his platoon.

When he reached the position, Inouye prepared to throw another grenade into the opening.

At that moment, an enemy soldier inside the bunker threw a grenade toward him.

The grenade exploded close to Inouye, **shattering his right arm**.

The blast left his arm badly wounded and nearly useless. But even in that moment, Inouye realized something dangerous—the grenade he had prepared to throw was still in his injured hand.

If it exploded there, it could kill him and the soldiers nearby.

Using his remaining strength, Inouye forced the grenade out of his damaged hand and threw it into the enemy bunker with his **left hand**.

The explosion destroyed the position.

Even after these terrible injuries, Inouye continued directing his men and firing his weapon at enemy positions on the ridge.

Only after the enemy resistance was finally broken did he allow other soldiers to carry him down the hillside for medical treatment.

His courage had helped his platoon capture the ridge and defeat the enemy defenses.

A Life-Changing Injury

Daniel Inouye's injuries were severe.

After the battle on the hillside ended, other soldiers carefully carried him down from the ridge so that medics could treat his wounds. The grenade explosion had caused terrible damage to his right arm, and doctors quickly realized that the injury could not be repaired.

In order to save his life, surgeons were forced to **amputate his right arm**.

Inouye spent many months recovering in military hospitals while doctors treated his other wounds and helped him regain his strength. The recovery process was long and difficult, but Inouye remained determined to rebuild his life.

Despite losing his arm, he refused to let the injury define him.

Instead, he focused on what he could still do and began planning for his future after the war. Like many soldiers who had been wounded in combat, Inouye carried the physical reminders of battle with him for the rest of his life.

But the courage he had shown on that hillside in Italy had already left a powerful impression on the soldiers who fought beside him.

They knew they had witnessed something extraordinary.

Daniel Inouye had continued fighting even after suffering terrible injuries, helping his men defeat the enemy defenses and capture the ridge.

He would go on to serve as the first Japanese American in both the House and Senate after his military career ended.

The Medal of Honor

Daniel Inouye's bravery on the hillside near **San Terenzo, Italy**, left a lasting impression on the soldiers who fought beside him.

They had seen him crawl toward enemy bunkers under heavy fire, destroy multiple machine-gun positions, and continue fighting even after being seriously wounded. His actions helped his platoon defeat the enemy defenses and capture the ridge.

For his courage, Inouye was originally awarded the **Distinguished Service Cross**, one of the highest military awards given by the United States.

Many years later, the U.S. military reviewed the records of Japanese-American soldiers who had served during World War II. Historians discovered that several acts of extraordinary bravery had not received the highest recognition they deserved.

As a result of that review, Daniel Inouye's award was **upgraded to the Medal of Honor**, the nation's highest military award.

On **June 21, 2000**, **President Bill Clinton** presented the Medal of Honor to Inouye during a ceremony at the **White House**.

The honor recognized the incredible courage he had shown during the battle in Italy more than fifty years earlier.

Daniel Inouye had proven that determination, leadership, and bravery could overcome even the most dangerous situations on the battlefield.

Why His Courage Still Matters

Daniel Inouye showed incredible courage during the battle in Italy.

Even after being shot and severely wounded by an exploding grenade, he refused to stop fighting until his soldiers had defeated the enemy defenses. His determination helped his platoon capture the ridge and complete their mission.

But Inouye's story did not end on the battlefield.

After recovering from his injuries, he returned to civilian life and continued serving his country in a different way. Inouye later became a leader in government and spent many years working to help improve the lives of the people he represented.

Throughout his life, he remained proud of the soldiers of the **442nd Regimental Combat Team**, the unit that proved the loyalty and courage of Japanese-American soldiers during World War II.

What kind of strength does it take to keep serving others even after facing such great hardship?

Daniel Inouye's story reminds us that courage is not only about a single moment of bravery.

Sometimes it is about continuing to serve others for an entire lifetime.

Because of his determination, leadership, and sacrifice, Daniel Inouye's legacy continues to inspire people today.

George Sakato

The Charge That Broke the Enemy Line

The soldiers of the **442nd Regimental Combat Team** moved slowly through the forests of eastern France in the fall of 1944. The air was cold, and the ground was covered with fallen leaves that crunched under their boots.

Somewhere ahead of them, German soldiers were waiting.

Suddenly, the quiet forest exploded with gunfire.

Enemy machine guns opened up from hidden positions along a hillside near the town of **Biffontaine, France**. Bullets tore through the trees as the American soldiers scrambled for cover behind rocks and fallen logs.

Several soldiers were hit.

Others were pinned down by the intense fire and could not move forward.

Among the men caught in the chaos was **Private George "Joe" Sakato**, a soldier in Company E of the 442nd Regimental Combat Team. Sakato and his fellow soldiers had already fought through many difficult battles in Europe.

But this attack was different.

Their platoon had lost its leader during the fight, and the men were stuck under heavy fire from enemy positions higher on the hill.

If no one acted quickly, the soldiers might remain trapped under enemy fire with no way forward.

As the gunfire continued to rip through the forest, Sakato realized that someone would have to take charge.

And in the next few moments, he would do something that changed the course of the battle.

Growing Up in America

George Taro Sakato was born on **February 19, 1921**, in **Colton, California**. His parents had come to the United States from Japan, hoping to build a better life for their family.

Like many children growing up in California at the time, Sakato spent his early years helping his family and going to school. He enjoyed sports and outdoor activities, and he learned the value of hard work from an early age.

But life was not always easy for Japanese-American families.

Even though Sakato had been born in the United States, many people still treated Japanese-Americans unfairly because of their background. These challenges became much worse after the United States entered **World War II** in 1941.

After the attack on **Pearl Harbor**, fear and suspicion spread across the country. The U.S. government ordered thousands of Japanese-American families living on the West Coast to leave their homes and move into special camps called **internment camps**.

George Sakato and his family were among those forced to leave their home.

They were sent to the **Poston War Relocation Center** in Arizona, where they lived behind barbed wire while the war continued.

George Sakato portrait

Despite this unfair treatment, Sakato still chose to serve his country.

Not long afterward, he joined the **U.S. Army** and became part of a special unit that would soon become famous for its courage.

The 442nd Goes to War

George Sakato joined the **442nd Regimental Combat Team**, a U.S. Army unit made up mostly of **Japanese-American soldiers**.

Many of these young men came from families who had been treated unfairly during the war. Some had relatives living behind barbed wire in internment camps while they themselves trained to fight for the United States.

Even so, they were determined to prove their loyalty.

The soldiers of the 442nd adopted a bold motto:

"Go for Broke."

It meant risking everything and holding nothing back.

The unit was eventually sent to fight in **Europe**, where the war looked very different from the battles in the Pacific. Instead of jungles and volcanic islands, soldiers faced forests, hills, villages, and cold weather.

The 442nd quickly earned a reputation for courage.

Again and again, its soldiers were sent into dangerous fighting and difficult missions. They became known for refusing to quit, even when the odds were against them.

By the fall of **1944**, Sakato and his fellow soldiers were fighting in the wooded hills of **eastern France**.

There, German defenders had taken strong positions in the forests and high ground.

The fighting was fierce, and every advance came with danger.

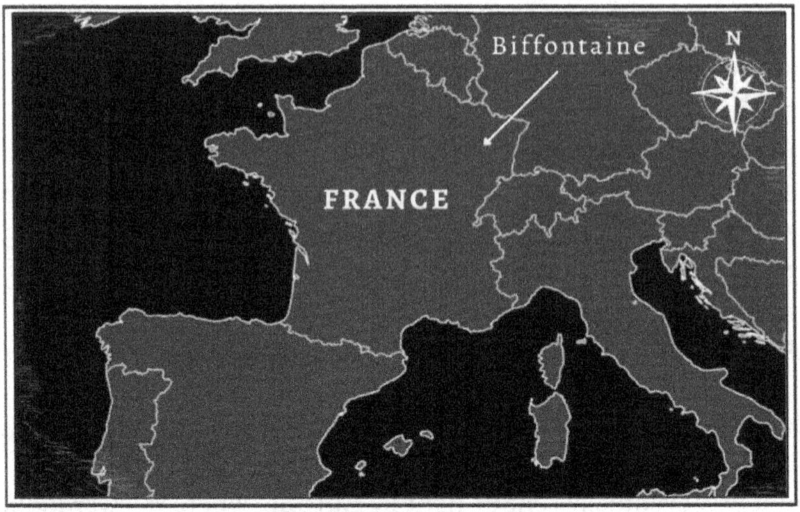

BIFFONTAINE: 1944

On **October 29, 1944**, during a battle near **Biffontaine, France**, George Sakato and his fellow soldiers would face one of the toughest moments of the war.

And on that hillside, Sakato would make a choice that helped turn the battle around.

Taking Charge

The attack had slowed under heavy enemy fire.

German machine guns hidden higher on the hillside kept the soldiers of the **442nd Regimental Combat Team** pinned down behind trees, rocks, and shallow patches of cover. Every time the Americans tried to move forward, another burst of bullets forced them back.

Then the situation grew even worse.

During the fighting, Sakato's **platoon leader was killed**.

For a moment, the men were left without their leader while enemy fire continued tearing through the forest around them.

The soldiers were still in danger, and the attack could not move forward unless someone acted quickly.

George Sakato understood that.

Instead of staying down, he rose up and began shouting for the other soldiers to follow him. With bullets still flying, Sakato charged uphill toward the German positions.

His sudden move caught the attention of the men around him.

Seeing Sakato rush forward, other soldiers jumped up and joined the attack.

The charge pushed the Americans closer to the enemy line and broke the moment of confusion that had followed the loss of their leader.

But the fighting was far from over.

German soldiers were still firing from strong positions on the hillside, and Sakato would soon find himself in the middle of a fierce close-range battle.

Charging the Enemy Line

George Sakato did not stop once he started up the hill.

With enemy bullets cutting through the trees around him, he charged toward the German positions that had pinned his platoon down. Other soldiers from the **442nd Regimental Combat Team** followed close behind, inspired by Sakato's decision to move forward.

The fight quickly turned into a close-range battle.

German soldiers fired from behind trees and hidden positions along the hillside. Sakato kept pressing ahead, moving through the enemy fire and firing his rifle as he advanced.

His sudden charge broke the grip that the German machine guns had held over the American soldiers.

As Sakato reached the enemy line, the battle became even more intense. He attacked the defenders at close range, helping drive them out of their positions and forcing them to fall back.

The soldiers behind him pushed forward as well.

What had begun as a platoon pinned down under heavy fire was now turning into an aggressive counterattack up the hillside.

The Germans began retreating.

But Sakato was not finished.

Instead of stopping at the first enemy positions, he kept moving forward with the attack. He chased the retreating German soldiers through the wooded hillside, determined not to let them regroup and open fire again.

By the end of the fighting, Sakato had helped destroy the enemy's defense on the hill and had played a major role in capturing several German soldiers.

His actions had changed the battle.

What could have become a deadly defeat for his platoon had turned into a hard-fought victory.

And George Sakato had led the charge that made it happen.

Turning the Battle Around

When the fighting on the hillside finally began to slow, the soldiers of the **442nd Regimental Combat Team** could see how much had changed.

Only a short time earlier, they had been pinned down under heavy enemy fire with no clear way forward. Their platoon leader had been killed, and the attack was in danger of falling apart.

Now the German line had been broken.

Because of George Sakato's charge, the soldiers of the 442nd had pushed uphill, overrun the enemy positions, and forced the defenders to retreat. In the confusion that followed, several German soldiers were captured.

The men who had fought beside Sakato knew what had made the difference.

When the attack had stalled and the danger was growing worse, Sakato had stood up and moved forward. His courage gave the other soldiers the push they needed to keep fighting.

What began as one soldier's decision quickly became a full assault that changed the outcome of the battle.

For the soldiers who were there, it was a moment they would never forget.

George Sakato had not waited for someone else to lead.

In one of the hardest moments of the fight, he had taken charge himself—and helped turn a desperate situation into a victory.

The Medal of Honor

George Sakato's bravery on the hillside near **Biffontaine, France**, left a lasting impression on the soldiers who fought beside him.

They had seen him stand up under heavy fire after the platoon leader was killed. They had watched him charge uphill, rally the men around him, and help drive the German defenders from their positions.

His actions had helped save the attack.

For his courage, Sakato was first awarded the **Distinguished Service Cross**, one of the nation's highest military honors.

But many years later, the U.S. military reviewed the records of Japanese-American soldiers who had served during World War II. Historians found that some acts of extraordinary bravery had not received the highest recognition they deserved.

As a result of that review, George Sakato's award was upgraded to the **Medal of Honor**, the highest military award given by the United States.

On **June 21, 2000, President Bill Clinton** presented the Medal of Honor to Sakato during a ceremony at the **White House**.

The honor recognized the courage he had shown on that hillside in France more than fifty years earlier.

George Sakato had proven that leadership does not always come from rank alone.

Sometimes it comes from the person willing to move first when everyone else is pinned down.

Why His Courage Still Matters

George Sakato's story reminds us that courage is not only about facing danger.

It is also about stepping forward when others need a leader.

On the hillside near **Biffontaine, France**, Sakato and his fellow soldiers were pinned down by enemy fire. Their platoon leader had been killed, and the attack was in danger of stopping.

In that moment, Sakato could have stayed where he was and waited for someone else to act.

Instead, he stood up and charged.

What kind of courage does it take to move first when everyone around you is under fire?

Sakato's actions helped his fellow soldiers find the strength to follow him. Because of his decision, the attack continued, the enemy line was broken, and lives were saved.

His story also reminds us of something else.

Even though Japanese-American families had been treated unfairly during the war, Sakato still chose to serve his country with honor and bravery.

That makes his courage even more powerful.

George Sakato's story shows that true leadership can come from anyone willing to act when it matters most.

Francis Currey
Stopping the German Tanks

The attack came at dawn.

Cold winter air hung over the village as American soldiers tried to hold their positions during the **Battle of the Bulge**, one of the fiercest battles fought in Europe during World War II.

Then the German tanks appeared.

Heavy armored vehicles rolled toward the American line, their engines growling as they pushed into the village streets. Behind them came enemy infantry, moving forward under the protection of the tanks.

The American soldiers were in serious danger.

If the tanks broke through, they could crush the defenders and open the way for a much larger German attack.

Among the soldiers caught in the fighting was **Private First Class Francis "Frank" Currey**.

Currey was an automatic rifleman, and he could see that the situation was quickly turning desperate. Enemy fire swept through the area, and American soldiers were being forced back as the tanks advanced.

But Currey did not run.

Instead, he moved through the village searching for a way to fight back.

Somewhere nearby, there had to be a weapon strong enough to stop a tank.

As shells exploded and bullets snapped through the cold air, Currey spotted what he needed.

A **bazooka**.

And in the next few moments, he would use it to face the enemy head-on.

A Soldier from New York

Francis Sherman Currey was born on **June 29, 1925**, in **Loch Sheldrake, New York**.

He grew up in a small town during the difficult years of the **Great Depression**, when many families had to work hard just to get by. Like many boys of his generation, Currey learned early how important determination and responsibility could be.

Friends and neighbors knew him as **"Frank."**

He was not famous, and he did not grow up expecting to become a war hero. He was simply a young American who came of age during a time when the world was at war.

After the United States entered **World War II**, millions of young men joined the military to serve their country.

Currey became a soldier in the **U.S. Army** and was assigned to an infantry unit that would eventually fight in some of the hardest battles in Europe.

Francis Currey portrait

Infantry soldiers faced danger at close range. They fought in towns, forests, fields, and roads where enemy fire could appear with little warning. Success often depended on quick thinking and courage under pressure.

By late **1944**, Currey was serving in **Belgium** as German forces launched a massive surprise attack through the Ardennes Forest.

The battle that followed would become one of the most famous of the war.

And in one snow-covered village, Frank Currey would face enemy tanks, rifle fire, and one of the most dangerous moments of his life.

The Battle of the Bulge

In **December 1944**, the war in Europe seemed to be turning in favor of the Allies.

American, British, and other Allied forces had pushed across France and into Belgium, driving German troops back toward their own borders. Many people hoped the war might soon be over.

But Germany launched one last major attack.

In the early days of winter, German forces struck through the **Ardennes Forest** in Belgium and Luxembourg, hoping to break through the Allied lines and force a change in the war.

The surprise attack created a huge bulge in the Allied front line.

That is why the battle became known as the **Battle of the Bulge**.

The fighting was brutal. Snow covered the ground, the air was bitterly cold, and soldiers often fought in forests and villages where visibility was poor and danger could appear suddenly.

German tanks played a major role in the attack.

If they broke through American defenses, they could overrun entire positions and open the way for larger enemy forces to advance.

That was the danger facing Frank Currey and the other American soldiers near **Malmedy, Belgium**, on **December 21, 1944**.

At dawn, German tanks and infantry began moving toward their position.

And the quiet village was about to become a battlefield.

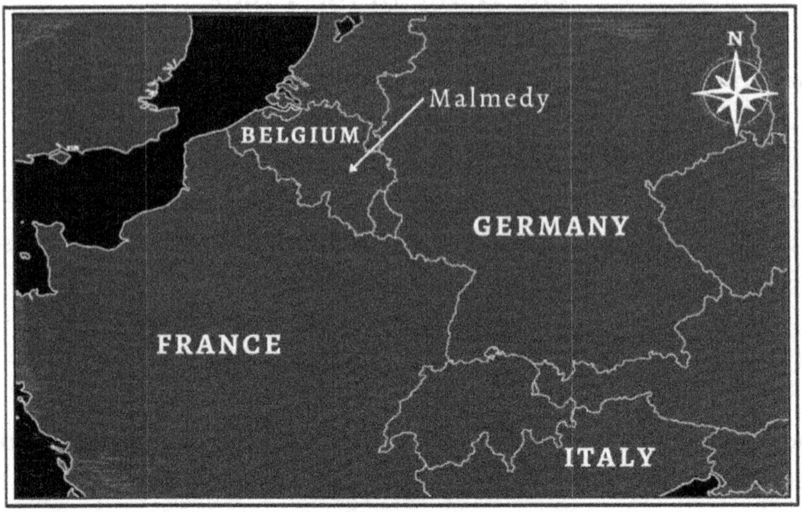

MALMEDY: 1944

Finding a Way to Fight

The German attack hit hard.

Tanks rolled into the area while enemy soldiers followed close behind, using the armored vehicles for cover. Shells exploded near the American positions, and rifle fire cracked through the village streets.

The defenders were being pushed back.

Frank Currey could see the danger immediately. If the tanks kept advancing, the American line might be broken before the soldiers could reorganize their defense.

Currey and the men around him needed something powerful enough to stop armored vehicles.

Moving through the chaos, he searched the area for weapons and ammunition that could still be used. Enemy fire swept the streets as he hurried from one position to another, looking for anything that could help.

Then he found it.

A **bazooka** and several rockets had been left behind near a vehicle.

Currey grabbed them and quickly moved into position.

He knew the bazooka gave him a chance—but only if he could get close enough and fire before the tank crews spotted him.

Keeping low, he moved through the village and waited for the right moment.

The tanks continued forward.

When one of them came into view, Currey raised the bazooka, aimed carefully, and prepared to fire.

In the middle of the freezing dawn battle, one young soldier was about to take on a German tank almost by himself.

Fighting Back in the Village

Frank Currey waited until the tank was close enough.

Then he fired.

The bazooka rocket streaked forward and hit the enemy vehicle, knocking it out and stopping its advance. For a moment, the German attack hesitated.

But the battle was far from over.

More enemy fire swept through the village, and additional German soldiers kept pushing forward behind the tanks. Currey knew that destroying one vehicle would not be enough to stop the attack.

So he kept moving.

As shells burst nearby and bullets snapped through the freezing air, Currey ran through the streets searching for better positions and more ways to fight back. He fired at enemy soldiers, changed locations, and refused to stay in one place long enough to become an easy target.

At one point, he saw several American soldiers trapped inside a house as German tanks and infantry closed in around them.

The men inside could not safely escape.

Currey realized they would be captured or killed if no one helped them.

Ignoring the danger, he moved toward the house under heavy fire. He gathered **anti-tank grenades** and used them against the enemy vehicles threatening the area. Then, from a nearby position, he kept firing at the advancing German soldiers to draw their attention away from the trapped Americans.

The plan worked.

The enemy was forced to focus on Currey instead of the soldiers in the house.

That gave the trapped men a chance to escape.

Still, Currey did not stop.

He climbed onto a half-track and used its machine gun to fire at the enemy, striking the German infantry and slowing their advance. When that position became too dangerous, he moved again, continuing to fire from different spots in the village.

Again and again, Currey put himself in danger so that others could survive.

By the end of the fighting, he had helped knock out enemy armor, rescue fellow soldiers, and hold back an attack that could have broken through the American line.

What made his actions so remarkable was not just one brave moment.

It was the way he kept adapting, moving, and fighting no matter how dangerous the situation became.

In the freezing chaos of that winter morning, Frank Currey became the kind of soldier every army needs in its hardest hour—calm, resourceful, and impossible to stop.

Holding the Line

By the time the fighting began to slow, the village was scarred by smoke, shattered buildings, and burning vehicles.

What had started as a sudden enemy attack had turned into a desperate fight for survival. German tanks had pushed forward at dawn, and infantry had followed close behind. For a time, it seemed possible that the American defense might be overrun.

But Frank Currey's actions helped change that.

By knocking out a tank, rescuing trapped soldiers, and repeatedly firing on the advancing enemy from different positions, he helped slow the attack and give the defenders time to hold their ground. The soldiers around him could see that his quick thinking and courage had made a real difference.

He had not simply stood and fired from one place.

He had moved wherever the danger was greatest, using whatever weapons he could find and putting himself at risk again and again to protect others.

The attack had been fierce, but the enemy had not broken through.

For the soldiers who fought there that morning, the battle was something they would never forget. And they would never forget the role Frank Currey had played in it.

In one of the hardest moments of the winter war in Europe, he had helped hold the line.

The Medal of Honor

Frank Currey's bravery during the fighting in Belgium did not go unnoticed.

The soldiers who had fought beside him reported how he had faced enemy tanks, attacked advancing infantry, and helped rescue trapped Americans while the battle was still raging around them. Again and again, he had moved into danger instead of away from it.

His actions had helped save lives and stop the enemy attack.

Francis Currey poses with his Medal of Honor

For his courage, **Private First Class Francis Sherman Currey** was awarded the **Medal of Honor**, the highest military award given by the United States.

On **July 27, 1945**, Major General Leland S. Hobbs presented the medal to Currey.

The award recognized more than a single brave act. It honored the way Currey kept thinking, adapting, and fighting during one of the most dangerous moments of the battle.

He had faced tanks, machine-gun fire, and advancing enemy soldiers, yet he refused to give up.

Frank Currey's story showed that battlefield courage is not only about strength.

Sometimes it is about staying calm in chaos, finding solutions under pressure, and continuing to fight when others are depending on you.

Why His Courage Still Matters

Frank Currey's story reminds us that courage is not always loud or dramatic at first.

Sometimes it begins with one person refusing to panic.

When enemy tanks rolled into the village and the situation grew more dangerous by the minute, Currey did not wait for someone else to solve the problem. He moved, searched for weapons, found new positions, and kept fighting wherever he was needed most.

That kind of bravery takes more than strength.

It takes clear thinking, determination, and the willingness to keep going when everything around you feels uncertain.

What kind of courage does it take to face tanks, enemy fire, and freezing winter conditions—and still keep helping others?

Frank Currey showed that heroes are not always the people with the highest rank.

Sometimes they are the ones who stay calm in chaos and do whatever is needed to protect the people beside them.

Because of his actions, lives were saved and the line held.

His story is a reminder that even in the hardest moments, one determined person can make a huge difference.

Audie Murphy

One Man Against an Army

The attack came fast.

Near **Holtzwihr, France**, German soldiers and tanks pushed toward the American line on **January 26, 1945**. The ground was cold and muddy, and the fighting quickly turned desperate as enemy forces advanced through the woods and open ground.

Among the American soldiers trying to hold their position was **Second Lieutenant Audie Murphy**.

Murphy commanded Company B, but the situation was growing worse by the minute. German tanks rolled forward while waves of enemy infantry followed behind them.

Then one of the American **tank destroyers** beside Murphy's position was hit.

Flames burst from the vehicle as it caught fire.

Most soldiers would have backed away from a burning armored vehicle, especially with enemy tanks and infantry closing in. At any moment, the fire could cause the ammunition inside to explode.

But Audie Murphy did something almost unbelievable.

He ordered his men to fall back to safer positions in the woods.

Then he stayed behind alone.

Murphy climbed onto the burning M10 tank destroyer and grabbed its **.50-caliber machine gun**. From that dangerous perch, with flames rising beside him and enemy fire all around, he began firing at the advancing Germans.

One man was now standing between the enemy attack and the soldiers behind him.

And he refused to give an inch of ground.

A Boy from Texas

Audie Leon Murphy was born on **June 20, 1924**, in **Hunt County, Texas**.

He grew up in a large family during the hard years of the **Great Depression**. Life was not easy. His family struggled for money, and from a young age Audie learned how to work hard, hunt for food, and help care for the people around him.

He was small for his age, but he was tough.

Murphy became an excellent shot as a boy, and he learned how to stay calm and focused in difficult situations. Those skills would later matter on the battlefield.

After the attack on **Pearl Harbor** in 1941, Murphy wanted to join the military and serve his country. But there was a problem.

He was still too young.

After his mother died, Murphy became even more determined to enlist. He tried to join different branches of the military before finally being accepted into the **U.S. Army**.

Audie Murphy portrait

Once he entered training, Murphy quickly proved that he was brave, disciplined, and willing to take on danger. Though he was young and not very large, he earned the respect of the soldiers around him.

By **1945**, Murphy was serving in **Company B, 1st Battalion, 15th Infantry Regiment, 3rd Infantry Division**.

He had already fought in some of the toughest battles in Europe.

But the fight near **Holtzwihr, France**, would become the most famous battle of his life.

Winter War in France

By **January 1945**, the war in Europe had entered a dangerous new phase.

The Allies had pushed across France and deep into enemy-held territory, but the fighting was far from over. German forces were still capable of launching powerful attacks, especially in the cold winter months when snow, mud, and freezing air made every movement harder.

In eastern France, American soldiers were trying to hold villages, roads, and wooded areas that were important to the Allied advance.

That was the situation near **Holtzwihr**.

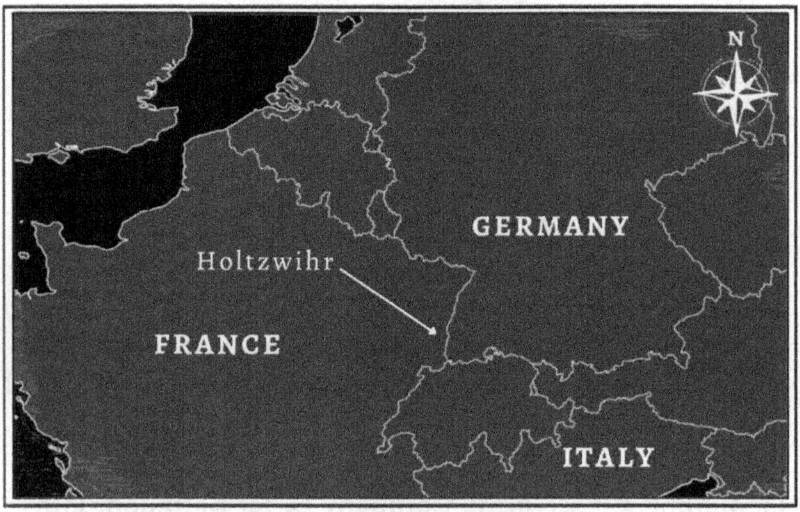

HOLTZWIHR: 1945

The area around the village was a mix of woods and open ground, making it possible for German tanks and infantry to attack together. If the enemy broke through, they could overrun American positions and force the defenders to retreat.

Murphy's company had already seen hard fighting.

On **January 26, 1945**, German forces launched a strong attack supported by **six tanks** and large numbers of infantry. The enemy pushed forward with enough force to threaten the entire American line.

The soldiers defending the area were badly outnumbered.

If the attack succeeded, Murphy's company could be surrounded and destroyed.

That was the danger facing Audie Murphy when the battle began.

And in the middle of that winter attack, he would make one of the boldest stands of the entire war.

The Line Begins to Break

The German attack hit with terrible force.

Tanks rolled forward across the frozen ground while enemy infantry advanced behind them, using the armored vehicles for cover. Shells burst near the American position, and machine-gun fire swept across the area.

Audie Murphy could see that his company was in grave danger.

The enemy had too many men, too many weapons, and too much momentum. If his soldiers stayed where they were, they could be overrun.

Murphy made a quick decision.

He ordered his men to **fall back into the woods** and take up more defensible positions. There, they would have a better chance to survive and continue fighting.

But while the others pulled back, Murphy stayed behind.

He knew that someone had to slow the German advance long enough for the rest of the company to escape.

Nearby stood an American **tank destroyer** that had been hit and set on fire. Flames climbed from the vehicle as smoke poured into the cold air.

It was a deadly place to stand.

At any moment, the burning vehicle could explode.

Even so, Murphy climbed onto it.

From the top of the flaming tank destroyer, he grabbed the **.50-caliber machine gun** and aimed at the advancing Germans.

Then he opened fire.

With enemy tanks moving toward him and infantry spreading across the field, Audie Murphy began fighting the attack almost entirely by himself.

Alone Against the Attack

Audie Murphy knew the burning tank destroyer could explode at any moment.

Still, he stayed on top of it and kept firing.

The **.50-caliber machine gun** in front of him was powerful enough to cut through advancing infantry, and Murphy used it with deadly accuracy. As German soldiers pushed forward across the open ground, he fired burst after burst into their ranks.

Enemy bullets struck around him.

The Germans could clearly see where the fire was coming from, and they answered with machine guns, rifles, and tank fire. Smoke from the burning vehicle rose around Murphy as flames climbed higher beside him.

But he refused to leave his position. .

Again and again, he fired at the advancing enemy, slowing the attack and forcing the German infantry to hesitate. The soldiers behind the tanks had expected to sweep forward quickly, but Murphy's gunfire kept cutting them down and breaking up their movement.

He was not fighting only with the machine gun.

While standing on the burning vehicle, Murphy also used the **field telephone** to direct American artillery fire onto the attacking Germans. Even as bullets snapped past him, he stayed in contact with artillery units, helping guide shells toward the enemy.

This made the attack even more difficult for the Germans.

Now they were facing not only Murphy's machine gun, but also exploding artillery shells falling around them.

The enemy kept pressing forward.

At one point, German soldiers got dangerously close to Murphy's position. According to his Medal of Honor citation, some came within **about ten yards** of where he stood. Even then, Murphy kept firing.

He was wounded in the leg during the fight, but he ignored the injury.

For nearly **an hour**, he held his ground alone.

The longer Murphy stayed on that burning tank destroyer, the more time he gave his company to reach safety and reorganize in the woods behind him. What could have become a complete disaster for the American soldiers was turning into a stubborn defense.

At last, the German attack began to falter.

The enemy infantry had taken heavy losses, and the advancing soldiers could not keep moving forward under such intense fire. Murphy's stand had broken their momentum.

Only then did he leave the burning vehicle and pull back to join his men.

But he did not stop fighting.

Once he reached the woods, he reorganized his company and led a counterattack that helped drive the Germans back.

In one of the most astonishing moments of World War II, Audie Murphy had stood alone against tanks and infantry, used a burning tank destroyer as a gun platform, directed artillery under fire, and refused to retreat until the enemy attack had been stopped.

Wounded but Still Leading

When Audie Murphy finally left the burning tank destroyer, the battle was not over.

He had been **wounded in the leg**, and the pain must have been intense. But Murphy did not let the injury stop him. Instead of collapsing or leaving the fight, he made his way back to the soldiers who had pulled into the woods.

There, he quickly took charge again.

The men of his company had survived because Murphy's stand had given them time to escape the German attack and form a new defensive line. Now they needed leadership for the next step.

Murphy gave it to them.

Even though he had just faced tanks and infantry almost alone, he reorganized his soldiers and prepared them to move forward again. Once the German attack had been slowed and broken apart, Murphy led his men in a **counterattack**.

The American soldiers pushed back against the enemy and regained ground that had nearly been lost.

For the men who fought beside him, Murphy's actions were unforgettable. He had not only held off the attack during its most dangerous moment, but had also returned to lead his company after being wounded.

That kind of courage is rare.

Audie Murphy had already saved his men once on the burning tank destroyer.

Now, by continuing to lead after his injury, he helped turn defense into victory.

The Medal of Honor

Audie Murphy's actions near **Holtzwihr, France**, were almost impossible to believe.

The soldiers who fought beside him had seen him stay behind alone, climb onto a burning tank destroyer, fire its machine gun at advancing German infantry, direct artillery under enemy fire, and then return to lead his company after being wounded.

His courage had helped save lives and stop a dangerous attack.

For those actions, **Second Lieutenant Audie Leon Murphy** was awarded the **Medal of Honor**, the highest military award given by the United States.

Audie Murphy portrait with his Medal of Honor

The ceremony took place on June 2, 1945, at an airfield near Werfen, Austria, where Lieutenant General Alexander M. Patch presented Murphy with the Medal of Honor.

The award recognized more than one brave moment.

It honored the way Murphy stayed calm under extreme danger, refused to retreat, and kept fighting even when the odds were against him.

By that time, Murphy was already known as a brave and effective soldier.

But his stand near Holtzwihr became the most famous action of his military career.

It showed the kind of courage that can change the outcome of a battle.

Audie Murphy's story would soon become known across the United States, and many Americans came to see him as one of the great heroes of World War II.

After surviving the war, he became a famous movie star, often playing himself in war films.

Why His Courage Still Matters

Audie Murphy's story reminds us that courage is not always about having the biggest weapon or the safest position.

Sometimes it means standing your ground when everything around you tells you to run.

Near Holtzwihr, Murphy faced enemy tanks, advancing infantry, and heavy fire. He was badly outnumbered, and the vehicle he stood on was already burning. Even so, he stayed in place long enough to protect his men and help stop the attack.

Then, after being wounded, he returned to lead them again.

What kind of courage does it take to do that?

Murphy's actions show that bravery is not the absence of fear. It is the decision to keep going when others are depending on you.

His story also reminds us that one person's determination can change the outcome of a battle. Because Murphy refused to give up, his company had time to survive, regroup, and fight back.

That is why his courage is still remembered today.

He showed that even in the hardest moment, one person can make an extraordinary difference.

Charles DeGlopper
Holding the Bridge Under Fire

Dawn had not yet fully broken when the soldiers moved forward.

Near the **Merderet River** in France on **June 9, 1944**, American troops were trying to reach a bridge that could help their forces continue pushing inland after the D-Day landings in Normandy.

But the road ahead was deadly.

German machine guns and rifles opened fire from hidden positions, cutting into the advancing Americans. Bullets tore through the morning air as the soldiers dropped for cover in ditches and along the roadside.

The attack was in trouble.

Men were being hit, and the platoon had become separated under the heavy fire. If the soldiers could not move forward, the bridge would remain in enemy hands.

Among the men caught in the attack was **Private First Class Charles DeGlopper**.

DeGlopper could see that the others needed time to get out of danger and reorganize. But with enemy fire pouring across the road, that would not happen unless someone stood up and fought back.

Then DeGlopper made his choice.

Instead of staying down, he rose from cover and opened fire on the German positions. His rifle and automatic fire cut through the chaos as he drew the enemy's attention toward himself.

The bullets kept coming.

But Charles DeGlopper kept firing.

And in the next few moments, his stand would help decide whether the men around him lived or died.

A Soldier from New York

Charles Nathan DeGlopper was born on **November 30, 1921**, in **Grand Island, New York**.

He grew up in a close-knit community near Buffalo, where family, faith, and hard work were important parts of daily life. Like many boys of his generation, he learned early how important it was to help others and do his part.

Friends and family remembered him as steady, dependable, and kind.

He was not the kind of person who searched for attention. Instead, he built a reputation as someone people could count on when things became difficult.

When the United States entered **World War II**, DeGlopper chose to serve his country.

He joined the **U.S. Army** and became part of the **325th Glider Infantry Regiment** of the **82nd Airborne Division**. Unlike paratroopers, glider infantry soldiers flew into battle in large wooden gliders that had no engines of their own once they were released from towing aircraft.

Charles DeGlopper portrait

It was a dangerous way to go to war.

These soldiers landed behind enemy lines carrying weapons and supplies, often arriving in darkness and confusion before having to fight immediately.

By June **1944**, DeGlopper and the other soldiers of the 82nd Airborne were in **France**, helping support the Allied invasion of Normandy.

There, on a road leading toward a bridge over the **Merderet River**, Charles DeGlopper would face the most important moment of his life.

Fighting in Normandy

By **June 1944**, the Allied invasion of **Normandy** was underway.

American, British, and Canadian forces had landed on the beaches of northern France on **D-Day**, hoping to break through German defenses and begin the long push toward victory in Europe.

But capturing the beaches was only the beginning.

To succeed, Allied forces also had to seize roads, bridges, and small towns farther inland. These routes were critical for moving soldiers, tanks, and supplies away from the coast.

That was why the area near the **Merderet River** mattered so much.

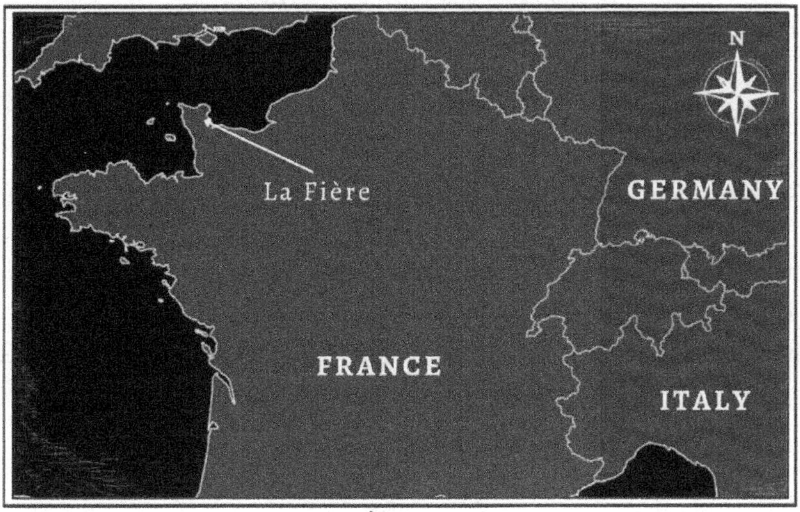

LA FIÈRE: 1944

Bridges over the river could help American troops continue advancing through the Normandy countryside. If German forces held those crossings, they could slow the Allied attack and keep reinforcements from moving forward.

The fighting inland was dangerous and confusing.

Hedgerows, narrow roads, fields, and ditches gave German soldiers many places to hide machine guns and rifles. A quiet road or field could suddenly become a deadly trap.

That was the danger facing the men of the **82nd Airborne Division** on **June 9, 1944**.

As they moved toward a bridge crossing, German defenders opened heavy fire and split the advancing platoon apart.

The road ahead became a killing zone.

And in that moment of chaos, Charles DeGlopper would make the choice that defined his story.

Pinned Down on the Road

The German fire was relentless.

Machine guns hammered the road while rifle shots cracked from hidden positions nearby. The American soldiers had been trying to push forward, but the sudden burst of enemy fire broke up their advance and forced many of them into ditches and shallow patches of cover.

Some men were hit.

Others were cut off from the rest of the platoon and could not safely move. The road toward the bridge had become too dangerous to cross.

Charles DeGlopper could see what was happening.

If the enemy kept firing at that pace, more soldiers would be wounded or killed before they could regroup. The men needed time to get off the exposed road and find better positions.

But time was exactly what they did not have.

Someone would have to stand up and draw the enemy's attention.

DeGlopper made his decision.

He rose from cover and began firing back at the German positions. His weapon cracked through the morning chaos as he stepped into the open, deliberately making himself the target so the others could move.

Bullets tore into the area around him.

Still, he kept shooting.

His brave stand gave the men around him a chance to crawl away from the road and pull back to safer ground.

But DeGlopper was only getting started.

Alone in the Open

Charles DeGlopper knew exactly what he was doing when he stepped into the road.

The German machine guns were already firing hard enough to tear apart the American attack. Men were pinned down in ditches, cut off from one another, and struggling to escape the deadly fire sweeping the area.

If no one distracted the enemy, more soldiers would be hit before they could get clear.

So DeGlopper made himself the target.

Standing in the open, he fired directly at the German positions while the men around him crawled and ran for better cover. His bursts of fire forced the enemy gunners to focus on him instead of the trapped Americans trying to pull back.

It was an almost impossible position to survive.

Bullets ripped through the air around him from several directions. German machine gunners and riflemen could clearly see where his fire was coming from, and they answered with everything they had.

But DeGlopper did not back away.

He kept firing.

According to his Medal of Honor citation, he moved through intense enemy fire and continued his attack even after being wounded. He advanced along the roadside, using his weapon to pour fire into the German positions and hold their attention as long as he could.

The enemy fire struck him again.

Still, he fought on.

Each extra second he remained standing gave the other soldiers more time to escape the killing zone and regroup. What DeGlopper was doing was not just brave. It was deliberate. He was sacrificing his own safety so that the men around him would have a chance to live.

The Germans kept shooting.

Even as he was hit multiple times, DeGlopper continued firing back, refusing to stop while any of his fellow soldiers were still in danger. His stand was so fierce that it pinned the enemy's attention on him and away from the others.

By the time his weapon finally fell silent, many of the American soldiers who had been trapped on the road had made it to safety.

Charles DeGlopper had given them that chance.

He was killed in the fight, but his actions had helped save the men around him and made it possible for the attack to continue.

On that dangerous road in Normandy, one soldier stood in the open against overwhelming fire so that others could survive.

And Charles DeGlopper never gave an inch.

Saving the Men Around Him

When the gunfire finally began to ease, the soldiers who had escaped the road understood what Charles DeGlopper had done for them.

Only moments earlier, they had been trapped under heavy enemy fire with little chance of moving safely. German machine guns had turned the road and the ditches beside it into a deadly trap.

Then DeGlopper stood up.

By drawing the enemy's attention to himself, he gave the others the time they needed to crawl away, find cover, and regroup. His stand had not stopped the danger completely, but it had changed the situation just enough to save lives.

The men who survived knew that many of them owed their lives to him.

DeGlopper had stayed in the open and kept firing even after being hit, refusing to stop while others were still in danger. His courage gave his fellow soldiers a chance they would not have had otherwise.

He did not live to see the battle continue.

But the example he set remained with the men who had fought beside him.

They had seen one soldier choose to face overwhelming fire so that others could live.

It was an act of sacrifice they would never forget.

The Medal of Honor

The bravery Charles DeGlopper showed on the road near the **Merderet River** was impossible for his fellow soldiers to forget.

They had seen him rise from cover while the platoon was pinned down under heavy German fire. They had watched him keep firing in the open, drawing the enemy's attention away from the trapped Americans trying to escape.

His actions had saved lives.

For that courage and sacrifice, **Private First Class Charles Nathan DeGlopper** was awarded the **Medal of Honor**, the highest military award given by the United States.

Because DeGlopper was killed in the battle, the award was presented **posthumously**, meaning after his death.

The honor recognized more than a single brave moment.

It recognized his willingness to face overwhelming danger so that the men around him could live. He knew the risk, yet he chose to stand and fight anyway.

Charles DeGlopper's story became one of the clearest examples of selfless courage during the fighting in Normandy.

He did not fight for recognition or glory.

He fought to protect the soldiers beside him.

And that is why his name is still remembered today.

Why His Courage Still Matters

Charles DeGlopper's story reminds us that courage is sometimes quiet until the moment it is needed most.

He was not fighting for fame. He was not trying to become a hero. He was a soldier who saw that the men around him were trapped under deadly fire and understood that someone had to act.

So he stood up.

What kind of courage does it take to face overwhelming danger, knowing you may not survive?

DeGlopper chose to risk everything so that others could live. He gave his fellow soldiers the time they needed to escape and regroup, even though doing so placed him directly in the enemy's path.

That kind of bravery is rare.

His story shows that true courage is not only about strength. It is also about sacrifice, loyalty, and the willingness to protect others when the cost may be very high.

Because of what Charles DeGlopper did on that road in Normandy, lives were saved.

And because of that choice, his example still speaks across the years.

John R. Fox
Calling Artillery on Himself

The town was under attack.

It was **December 26, 1944**, and German soldiers were pushing into the Italian village of **Sommocolonia**. The streets and stone buildings that had been quiet only hours before were now filled with gunfire, explosions, and shouting.

American defenders were trying to hold on.

But the enemy attack was strong, and German troops were moving deeper into the town. One by one, the defenders were forced to fall back as the fighting grew more intense.

Among the men still in position was **First Lieutenant John R. Fox**, an artillery officer.

Fox's job was not to fire a rifle from the front line. His job was to observe the battlefield and call in **artillery fire**—powerful shells launched from far away onto enemy positions.

From inside a house near the center of the fighting, Fox could see the danger clearly.

German soldiers were closing in.

If they broke through completely, they could overwhelm the remaining defenders and take control of the town before help could arrive.

Fox knew what that meant.

The enemy had come so close that ordinary artillery fire would no longer be enough.

If he wanted to stop the attack, he would have to order the shells to land almost directly on his own position.

And that is exactly what John Fox prepared to do.

A Teacher from Ohio

John Robert Fox was born on **May 18, 1915**, in **Cincinnati, Ohio**.

He grew up in a time when education, discipline, and hard work were deeply valued. Fox was known as a bright and thoughtful young man, and he eventually attended **Wilberforce University**, a historically Black college in Ohio.

Before joining the Army, Fox worked as a **teacher**.

That fact makes his story especially powerful. He was not someone who grew up dreaming only of battle. He was an educated man who cared about learning and helping others build a better future.

But when the United States entered **World War II**, Fox chose to serve his country.

He joined the **U.S. Army** and became part of the **92nd Infantry Division**, one of the major African American divisions that fought in the war.

John R. Fox portrait

Fox became an **artillery officer**, a dangerous and important role. Artillery officers had to move close enough to the front to see enemy positions clearly, then call in shellfire from American guns positioned farther away.

It was a job that required calm thinking under enormous pressure.

And in the mountains of Italy, John Fox would show exactly that.

Winter Fighting in Italy

By the end of **1944**, the war in Italy had become a hard and exhausting struggle.

Allied forces had already fought their way north through mountains, valleys, and heavily defended towns. But German troops continued to resist, using the rugged landscape to slow every advance.

The fighting was especially difficult in winter.

Cold weather, steep hills, and narrow village streets made it hard for soldiers to move and even harder to attack. A small town on high ground could become an important strongpoint if enemy soldiers used its buildings and roads for defense.

That was the situation at **Sommocolonia**, a hill town in the **Serchio River Valley**.

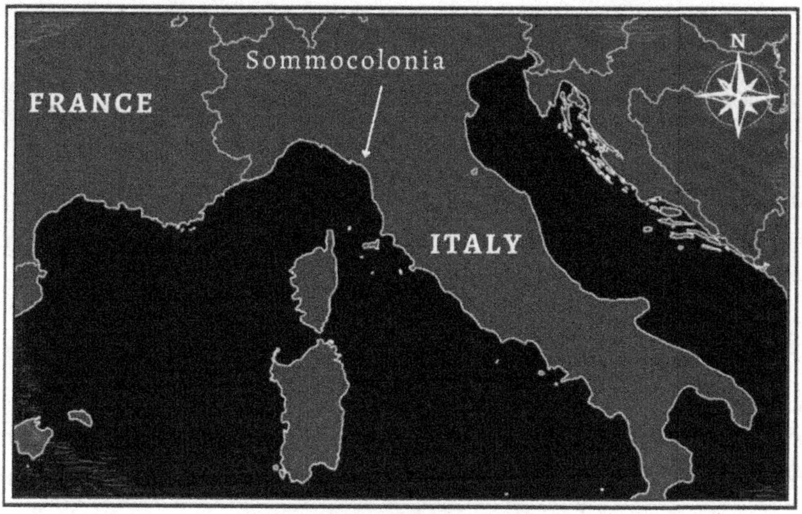

SOMMOCOLONIA: 1944

The village sat in a position that mattered to both sides. If German forces captured it, they could threaten nearby Allied positions and make it harder for American troops to hold the surrounding area.

On **Christmas Day and the day after**, German forces launched a strong attack against the town.

The fighting quickly became fierce.

American defenders tried to hold the streets and houses, but the enemy kept pressing forward. As more German soldiers entered the village, the defenders were forced back step by step.

Among those still in position was **John R. Fox**, serving as a forward observer for American artillery.

As the attack grew more dangerous, Fox could see that the enemy was getting closer and closer.

And soon, he would face a decision that almost no soldier ever has to make.

The Enemy Closes In

John Fox stayed at his observation post while the fighting moved closer.

From inside a house near the center of town, he watched German soldiers push deeper into the streets of **Sommocolonia**. The American defenders were being forced back, and the enemy was no longer far away.

They were almost on top of him.

Fox's job was to report what he saw and call for artillery fire on enemy positions. Earlier in the battle, that had meant directing shells onto streets and approaches outside the town.

But now the situation had changed.

German troops were moving so close to Fox's position that ordinary artillery adjustments would no longer stop them. If the shells landed too far away, the enemy would keep advancing. If they landed where they needed to, the blasts would strike dangerously close to Fox himself.

Still, Fox did not leave.

He stayed on the radio and continued calling in artillery, adjusting the fire as the Germans moved nearer and nearer. The men around him could hear the danger in every report. Enemy soldiers were entering the area, and there was almost no space left between Fox and the attack.

At last, Fox understood what had to happen.

If the town was going to be held long enough for others to regroup, the artillery would have to come directly onto his own position.

And John Fox made the decision to call it in.

Bring It In Closer

John Fox picked up the radio and gave the order.

American artillery had already been firing on the German attack, but now Fox told the gunners to bring the shells **closer**.

The enemy was no longer at the edge of town.

They were almost on top of him.

From his position inside the house, Fox could see German soldiers moving through the streets and between the buildings. The American defenders nearby had been forced back, and the enemy was closing in fast.

If the artillery stayed too far out, it would not stop them.

So Fox adjusted the fire again.

Closer.

The soldiers operating the artillery understood what that meant. The shells were now falling dangerously near Fox's own location. One officer warned him that the rounds were getting too close.

But Fox did not change his order.

He knew exactly what he was doing.

The German attack had gained too much ground, and there was only one way left to stop it. If the shells landed where they needed to, they would strike not only the enemy soldiers in the streets below, but also the building where Fox was still observing.

Even so, he remained at his post.

Again he called in the artillery.

Closer still.

Shells screamed down into the town. Explosions ripped through the streets and around the houses as the barrage struck the advancing Germans. Stone walls shattered. Debris flew through the air. The attack that had seemed unstoppable was suddenly caught in a storm of American fire.

Still the enemy pushed forward.

Fox made one final adjustment.

This time, he ordered the shells to fall directly on his own position.

It was the last and most dangerous choice he could make.

He knew that if the artillery struck where he was standing, he would almost certainly be killed. But he also knew that if he did not make that call, the German soldiers would break through and overwhelm the remaining defenders.

So he stayed on the radio and gave the order anyway. His final command was "*Fire it! There's more of them than there are of us. Give them hell!*"

Moments later, the artillery crashed down on the area.

The explosions tore through the enemy troops who had closed in around Fox's position. The barrage was devastating. German soldiers were killed, the attack was broken apart, and the enemy advance was stopped long enough for American forces to regroup and fight back.

John Fox was killed in the bombardment.

But his decision had made all the difference.

By calling artillery fire onto his own position, he sacrificed his life to stop the attack and save the soldiers around him.

The Town Was Saved

When the artillery fire finally stopped, the German attack had been shattered.

The enemy soldiers who had pushed deep into **Sommocolonia** had been hit hard by the barrage that John Fox called down on his own position. What had seemed only moments earlier like a successful breakthrough was suddenly thrown into chaos.

The Germans had lost their momentum.

That gave the American defenders something they desperately needed: **time**.

With the attack broken apart, other U.S. forces were able to regroup and fight back. The town had not been lost without a fight, and Fox's final decision helped prevent the enemy from taking complete control at that critical moment.

The soldiers who later learned what had happened understood the cost.

John Fox had stayed at his post while the enemy closed in. He had continued adjusting artillery fire even when he knew the shells were falling closer and closer to him. And when the moment came, he chose to sacrifice himself rather than let the attack succeed.

His action saved lives.

His body was eventually found with approximately 100 dead German soldiers surrounding his position, proving the efficacy of his sacrifice.

It also showed the kind of courage that is hard to imagine—a willingness to give up everything so that others might survive and the mission might continue.

The teacher from Ohio had made his final lesson one of duty, sacrifice, and extraordinary bravery.

The Medal of Honor

The bravery John Fox showed in the defense of **Sommocolonia** was remembered by the soldiers who survived the battle.

They knew that when the enemy closed in, Fox had not abandoned his post. Instead, he stayed on the radio and called in artillery fire closer and closer until the shells landed directly on his own position.

His sacrifice helped stop the German attack and gave American forces the chance to regroup.

For many years, however, Fox did not receive the nation's highest military award.

Like several other African American soldiers of World War II, his actions were later reviewed as historians and military leaders took a closer look at acts of bravery that had not been fully recognized at the time.

As a result of that review, **First Lieutenant John Robert Fox** was awarded the **Medal of Honor**, the highest military award given by the United States.

On **January 13, 1997**, the medal was presented to his widow, **Arlene C. Fox**, during a White House ceremony led by **President Bill Clinton**.

The award honored a man who had given his life to stop the enemy and protect the soldiers around him.

John Fox's courage had not been forgotten.

And many years after the battle, the nation finally gave his sacrifice the recognition it deserved.

Why His Courage Still Matters

John Fox's story reminds us that courage is not always about charging forward with a weapon in your hands.

Sometimes it means staying at your post when every second grows more dangerous.

Fox was a teacher before he became a soldier. He chose a job in the Army that required calm thinking, clear judgment, and the ability to make life-or-death decisions under pressure. In the final moments of the battle, he used those skills in the most selfless way possible.

He knew what would happen if he called the artillery onto his own position.

And he made the call anyway.

What kind of courage does it take to sacrifice your own life so that others can survive?

John Fox's actions show that true bravery can come from duty, sacrifice, and love for the people beside you.

Because he refused to abandon his post, the enemy attack was broken and lives were saved.

His story is still remembered because it shows the very highest kind of courage: choosing to give everything so that others may live.

Harold Gonsalves
The Grenade That Saved His Squad

The hillside shook with explosions.

On **April 15, 1945**, American Marines were fighting their way across **Okinawa Shima**, one of the hardest battlefields of World War II. Enemy shells burst across the rocky ground while rifle and machine-gun fire snapped through the air.

The Marines were pushing toward strong Japanese defenses.

Among them was **Private First Class Harold Gonsalves**, serving with the **6th Marine Division**.

The fighting was fierce and close. Japanese defenders had built strong positions into the high ground, and every step forward brought more danger. Marines had to move carefully, using craters, rocks, and broken ground for cover as they advanced.

Gonsalves was part of a small group moving near the front of the attack.

Then, in an instant, everything changed.

A **Japanese grenade** landed close to the Marines.

There was almost no time to react.

The men nearby knew the grenade would explode within seconds. If it did, the blast would send deadly fragments through the group.

In that brief, terrible moment, Harold Gonsalves made his choice.

A Young Marine from California

Harold Gonsalves was born on **January 28, 1926**, in **Alameda, California**.

He grew up during the difficult years of the **Great Depression**, when many families had to work hard and make do with very little. Like many boys of his generation, Harold learned early the value of responsibility, toughness, and helping the people around him.

Friends and family remembered him as energetic and dependable. He was still very young when the world was pulled into war, but like many other teenagers of the time, he soon felt the call to serve.

After the United States entered **World War II**, Gonsalves joined the **U.S. Marine Corps Reserve**.

He became part of a generation of young Americans who left home while still in their teens to fight in battles far from anything they had ever known. In the Marines, he trained hard and learned to endure long marches, rough conditions, and dangerous combat.

By **1945**, Gonsalves was serving with **Battery L, 4th Battalion, 15th Marines, 6th Marine Division**.

Harold Gonsalves portrait

That unit would soon take part in one of the toughest campaigns of the entire Pacific war.

The battle for **Okinawa**.

And there, on a shattered hillside under enemy fire, Harold Gonsalves would face the most important moment of his life.

The Battle for Okinawa

By the spring of **1945**, the war in the Pacific had entered its final and most brutal stage.

American forces had fought their way across the ocean for years, capturing island after island as they moved closer to **Japan**. Each new battle had been costly, but military leaders believed the next target was especially important.

That target was **Okinawa**.

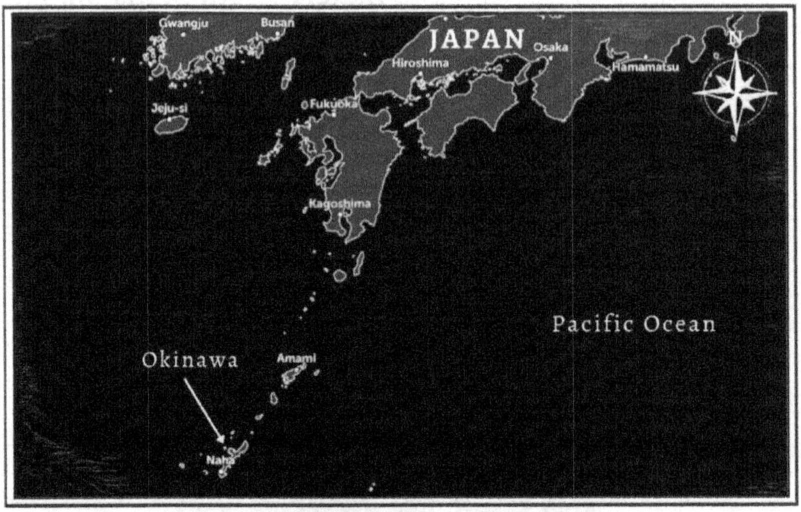

OKINAWA: 1945

The island lay close enough to Japan to serve as a major base for future air attacks and military operations. If American forces could capture it, they would have a much stronger position for the final phase of the war.

But the Japanese defenders were ready.

They had built strong defensive positions across the island, using ridges, hills, caves, and fortified bunkers to slow the American advance. Instead of trying to stop the invasion only at the beaches, they prepared to fight a long, deadly battle inland.

That made every advance dangerous.

Marines and soldiers often had to move uphill under artillery, mortar, rifle, and machine-gun fire. Even a small rise in the ground could hide enemy positions strong enough to stop an entire attack.

That was the kind of battlefield Harold Gonsalves faced on **April 15, 1945**.

His unit was moving against a strongly defended Japanese position on the island.

And in the middle of that fierce attack, a single grenade would turn one terrible moment into a story of lasting courage.

Moving Up the Hill

The Marines kept pushing forward.

Enemy fire came from high ground ahead, where Japanese defenders had built strong positions into the hillside. Mortar rounds burst nearby, and rifle shots cracked through the air as the attack moved uphill.

Harold Gonsalves was advancing with a small group near the front.

The terrain was rough and exposed, with little cover beyond rocks, shallow depressions, and the uneven slope of the hill itself. Every few steps forward felt dangerous. The Marines knew the enemy was watching and waiting.

Still, they had to keep moving.

The attack could not succeed unless the men on the front line got close enough to locate the hidden positions and help direct fire onto them. Gonsalves and the Marines around him pressed ahead through the chaos, staying low as shells and bullets tore across the hillside.

Then the danger came even closer.

A Japanese grenade landed near the group.

It fell within only a few feet of the Marines, close enough that there was almost no time to think and almost no room to escape. Everyone nearby knew what would happen next.

The blast would come in seconds.

And on that exposed hillside, with the enemy fire still crashing around them, Harold Gonsalves had only an instant to decide what to do.

A Split-Second Sacrifice

The grenade landed in the middle of the group.

There was no time to throw it back. No time to run. No time even to shout more than a warning.

Harold Gonsalves understood that immediately.

In combat, a grenade could explode in only a few seconds, sending metal fragments in every direction. On that exposed hillside, with the Marines crowded close together and enemy fire still pouring in, the blast could kill or badly wound several men at once.

Gonsalves made his decision in an instant.

He threw himself onto the grenade.

A moment later, it exploded.

The force of the blast struck Gonsalves directly, but his body absorbed the explosion and shielded the Marines beside him from the worst of the deadly fragments. Men who might have been killed or terribly wounded survived because of what he did.

His action lasted only a second.

But that second changed everything for the soldiers around him.

The battle did not stop. Enemy fire still swept the hillside, and the Marines still had to keep moving under dangerous conditions. Yet the men who had just witnessed Gonsalves's sacrifice knew exactly what had happened.

He had given his life to save theirs.

In war, courage can appear in many forms. Sometimes it is a long fight against impossible odds. Sometimes it is a steady act of leadership under fire. And sometimes it is one sudden choice made with no time to hesitate.

That is what Harold Gonsalves did on Okinawa.

He did not wait for someone else to act. He did not think first about his own safety. He saw the danger facing the men beside him and chose to protect them, even though the cost would be his own life.

The Marines around him survived the blast and continued the fight.

Because of Gonsalves's action, the attack did not lose more men in that terrible moment. His sacrifice spared others and allowed the mission to continue.

What happened on that hillside was heartbreaking.

But it was also unforgettable.

In one of the fiercest battles of World War II, a young Marine from California made a choice of extraordinary courage. He gave everything he had so that the men beside him could live.

The Men He Saved

The Marines beside Harold Gonsalves survived.

Only moments earlier, they had been exposed on the hillside with enemy fire coming down around them. If the grenade had exploded where it landed, several of them could have been killed or badly wounded.

Instead, Gonsalves took the full force of the blast himself.

The men around him understood exactly what he had done. In the middle of battle, with almost no time to think, he had made a choice to save others instead of himself.

That kind of sacrifice is hard to imagine.

The fighting on Okinawa continued, but the Marines who lived through that moment would never forget it. They kept moving under fire and continued the attack, carrying with them the memory of the young Marine who had given his life for theirs.

Gonsalves did not survive to see the battle end.

But his actions mattered immediately. He had preserved the strength of the group, protected the men closest to him, and made it possible for the attack to continue without even greater loss.

His courage lasted only an instant.

Its impact lasted much longer.

The Medal of Honor

The sacrifice Harold Gonsalves made on the hillside in Okinawa was never forgotten by the Marines who were there.

They knew that when the grenade landed among them, he had only seconds to act. By throwing himself onto it, he gave his own life to save the men beside him.

For that extraordinary courage, **Private First Class Harold Gonsalves** was awarded the **Medal of Honor**, the highest military award given by the United States.

Because he was killed in the action, the medal was awarded **posthumously**, meaning after his death.

The honor recognized a moment of complete selflessness. Gonsalves had no time to plan and no chance to protect himself. He saw the danger, understood what it meant for the others around him, and chose to sacrifice everything to save them.

His action lasted only an instant, but it showed the highest kind of bravery.

Harold Gonsalves did not survive the battle for Okinawa.

But his courage saved lives, and that is why his name is still remembered among the heroes of World War II.

Why His Courage Still Matters

Harold Gonsalves's story brings together many of the ideas that run through this book.

He was young. He was far from home. He was fighting in one of the hardest battles of World War II. And when a terrible moment came with almost no time to think, he chose to protect the men beside him.

That kind of courage is hard to measure.

He did not have time to make a speech. He did not have time to plan. He had only a second or two to decide what mattered most.

And he chose others.

Throughout this book, you have read about soldiers, Marines, medics, pilots, and officers who faced danger in many different ways. Some fought from the air. Some charged bunkers. Some held off enemy attacks. Some gave their lives so others could live.

Harold Gonsalves's story reminds us that courage can appear in a single instant.

Sometimes bravery lasts for hours in the middle of battle. Sometimes it is shown through leadership, endurance, or skill. And sometimes it is revealed in one sudden choice that changes everything for the people nearby.

What kind of person makes a decision like that?

A person who values the lives of others. A person who is loyal. A person who understands that some things are worth more than personal safety.

That is why Harold Gonsalves's story still matters.

His sacrifice saved lives on that hillside in Okinawa. But it also leaves behind something more: an example of selflessness that can still inspire readers many years later.

As this book comes to an end, the stories of these Medal of Honor heroes remind us that courage is not just part of history.

It is a choice people make.

Again and again, in different places and under different kinds of pressure, these men chose duty, sacrifice, and bravery.

And because they did, their stories are still worth telling.

Think Like a Hero

The men in this book faced danger in very different ways.

Some charged enemy positions. Some rescued the wounded. Some led others under fire. Some gave their lives so others could live.

Their stories are part of history, but they can also make us think about our own choices.

You may never face the kind of battles these heroes faced. But you will face moments when courage, loyalty, leadership, and selflessness matter.

Use these questions and challenges to think more deeply about what real courage looks like.

Courage Questions

1. What does courage mean to you?
2. Do you think someone can be brave even if they feel afraid? Why?
3. Which hero in this book showed the kind of courage you admire most?
4. Was there a story in this book that surprised you? What made it surprising?
5. Which hero do you think stayed the calmest under pressure?
6. Which story felt the most emotional to you? Why?
7. Do you think courage usually happens in one big moment, or in many smaller choices?
8. What is the difference between bravery and recklessness?
9. Why do you think so many of these heroes chose to protect others instead of saving themselves?
10. Which hero do you think showed the strongest leadership?

Real-Life Hero Thinking

11. Can someone be a hero even if no one ever knows what they did?
12. What are some ways a person can show courage in everyday life?
13. When is it hardest to do the right thing?
14. Have you ever seen someone stand up for another person? What happened?
15. What does loyalty mean?
16. Why is protecting others such an important kind of courage?
17. Do you think courage looks the same in school, at home, on a team, and in battle?
18. Which quality matters most in a hero: bravery, selflessness, leadership, discipline, or kindness?
19. If you were making a list called "Signs of a Real Hero," what would you include?
20. Why is it important to remember true stories of courage from history?

Hero Challenges

21. Pick one hero from this book and describe the hardest choice he had to make.
22. Write about a time when you had to do something difficult even though you were nervous.
23. Choose three heroes from the book. What did they all have in common?
24. Which story in this book would you most want to tell a friend or family member? Why?
25. Imagine you could ask one hero from this book a question. Who would you ask, and what would you want to know?
26. Make a list of five words that describe a hero.
27. Which story best shows that courage is about helping others?
28. Which hero do you think showed the most determination?
29. Think of a person in your own life who is brave in a quieter way. What makes that person brave?
30. After reading this book, has your idea of a hero changed? How?

Test Your Knowledge

You've read about medics, Marines, pilots, soldiers, and officers who showed extraordinary courage during World War II.

Now see how much you remember.

Some questions ask about facts. Others ask you to remember which hero did what. Take your time and do your best.

Part 1: Multiple Choice

1. Which hero was a combat medic who refused to carry a weapon?
   A. Audie Murphy
   B. Desmond Doss
   C. Jack Lucas
   D. Richard Bong
2. Which hero threw himself onto grenades to save the Marines beside him?
   A. Jack Lucas
   B. George Sakato
   C. Francis Currey
   D. John Basilone
3. Which hero became America's top fighter ace in World War II?
   A. Richard Bong
   B. Henry Erwin
   C. Daniel Inouye
   D. Charles DeGlopper
4. Which hero carried a burning phosphorus bomb through a bomber to save his crew?
   A. Audie Murphy
   B. Henry Erwin
   C. Rodger Young
   D. John R. Fox

5. Which hero climbed onto a burning tank destroyer and fired its machine gun?
   A. Jack Lummus
   B. John Basilone
   C. Audie Murphy
   D. Harold Gonsalves
6. Which hero called artillery fire onto his own position to stop the enemy?
   A. John R. Fox
   B. George Sakato
   C. Charles DeGlopper
   D. Daniel Inouye
7. Which hero's battle took place on the island of Iwo Jima?
   A. Desmond Doss
   B. Jack Lucas
   C. John R. Fox
   D. Francis Currey
8. Which hero was known for destroying enemy bunkers with a flamethrower?
   A. Hershel "Woody" Williams
   B. Rodger Young
   C. Jack Lummus
   D. Harold Gonsalves
9. Which hero served in the famous 442nd Regimental Combat Team?
   A. Richard Bong
   B. Daniel Inouye
   C. Henry Erwin
   D. John Basilone
10. Which hero's story took place during the Battle of the Bulge?
    A. Francis Currey
    B. Harold Gonsalves
    C. Desmond Doss
    D. Charles DeGlopper

Part 2: Match the Hero

Write the correct hero's name next to each description.

11. ________ stayed behind on Okinawa and rescued about 75 wounded soldiers.
12. ________ held the line on Guadalcanal with machine guns.
13. ________ charged a machine-gun nest after being wounded and told others, "Don't give up."
14. ________ was a teacher before the war and later called artillery on his own position.
15. ________ led his men forward on Iwo Jima even after stepping on a land mine.
16. ________ stood in the open under heavy fire so the men around him could escape.
17. ________ led a charge after his platoon leader was killed in battle.
18. ________ saved the Marines beside him by throwing himself onto a grenade on Okinawa.

Part 3: Short Answer

19. What was the Medal of Honor awarded for?
20. Why was Okinawa such an important battle in the Pacific war?
21. What was special about the 442nd Regimental Combat Team?
22. Why were bunkers and hidden defensive positions so dangerous in battles like Iwo Jima and Italy?
23. Name one hero in this book who survived the action that earned him the Medal of Honor.
24. Name one hero in this book who was killed during the action for which he received the Medal of Honor.
25. Which story in the book do you think would be hardest to forget, and why?

*Answer key available in the back of this book.

Afterword

Why These Stories Matter

Thank you for spending time reading these stories.

We hope this book introduced you to men whose courage, sacrifice, and character will stay with you long after you finish the final page. The battles in this book took place many years ago, in places far from most of our everyday lives. But the choices these men made still matter.

Each hero in this book faced danger in a different way. Some charged enemy positions. Some rescued the wounded. Some protected the men beside them. Some gave their lives so that others could live. Their stories remind us that bravery can take many forms, but it always begins with a choice.

That choice is what makes these men worth remembering.

Not every war is easy to understand. Not every battle is simple to explain. And history is full of moments when leaders, nations, and armies made decisions that still spark debate today. But even in the middle of difficult and tragic wars, there are still people whose courage deserves to be honored.

That is especially true of the men in this book.

These were men who chose duty over comfort, sacrifice over safety, and service over self-interest. Many of them were young. Many of them came from ordinary homes and ordinary communities. Yet in extraordinary moments, they revealed the kind of character that makes a country stronger.

And when men like that are lost, it is a loss for all of us.

Their value was not only in what they did in battle. Men of courage like these go on to shape families, neighborhoods, schools, churches, teams, and communities. They become fathers, mentors, workers, leaders, and examples to others. Even those who died young left behind a legacy much larger than a single battlefield.

That is one reason these stories matter so much.

At Good Buddy Books, we believe history is not just about dates, places, and battles. It is also about people. It is about the values they lived by, the decisions they made, and the examples they leave behind. We hope this book helped bring those stories to life in a way that felt exciting, meaningful, and worth remembering.

More than anything, we hope these stories encourage readers to think deeply about courage.

What does it mean to protect others?
What does it mean to lead?
What does it mean to stay loyal, steady, and brave when the cost is high?

Those questions are not only for soldiers. They are for all of us.

Thank you for spending time with these heroes and the history they were part of. We hope this book helped you see that courage is not only something to admire in the past. It is something to practice in the present.

Warmly,

The Good Buddy Books Team

Research Sources

Desmond Doss
Congressional Medal of Honor Society
U.S. Army Center of Military History – Medal of Honor Citations
National WWII Museum – Battle of Okinawa

John Basilone
Congressional Medal of Honor Society
U.S. Marine Corps History Division
National WWII Museum – Guadalcanal Campaign

Rodger Young
Congressional Medal of Honor Society
U.S. Army Center of Military History – Medal of Honor Citations
National WWII Museum – Solomon Islands Campaign

Henry "Red" Erwin
Congressional Medal of Honor Society
U.S. Army Center of Military History – Medal of Honor Citations
U.S. Air Force Historical Research Agency
National WWII Museum – Strategic Bombing Campaign

Jack Lucas
Congressional Medal of Honor Society
U.S. Army Center of Military History – Medal of Honor Citations
National WWII Museum – Battle of Iwo Jima
U.S. Marine Corps History Division

Jack Lummus
Congressional Medal of Honor Society
U.S. Army Center of Military History – Medal of Honor Citations
National WWII Museum – Battle of Iwo Jima
U.S. Marine Corps History Division

Hershel "Woody" Williams
Congressional Medal of Honor Society
U.S. Army Center of Military History – Medal of Honor Citations

National WWII Museum – Battle of Iwo Jima
U.S. Marine Corps History Division

Richard Bong
Congressional Medal of Honor Society
U.S. Army Center of Military History – Medal of Honor Citations
National WWII Museum – Air War in the Pacific
Smithsonian National Air and Space Museum
U.S. Air Force Historical Research Agency

Daniel Inouye
Congressional Medal of Honor Society
U.S. Army Center of Military History – Medal of Honor Citations
National WWII Museum – Italian Campaign
U.S. Army Center of Military History – 442nd Regimental Combat Team
Smithsonian National Museum of American History

George Sakato
Congressional Medal of Honor Society
U.S. Army Center of Military History – Medal of Honor Citations
National WWII Museum – European Theater
U.S. Army Center of Military History – 442nd Regimental Combat Team
Densho Encyclopedia

Francis Currey
Congressional Medal of Honor Society
U.S. Army Center of Military History – Medal of Honor Citations
National WWII Museum – Battle of the Bulge
U.S. Army Center of Military History – Ardennes-Alsace Campaign
The National Archives

Audie Murphy
Congressional Medal of Honor Society
U.S. Army Center of Military History – Medal of Honor Citations
National WWII Museum – European Theater
U.S. Army Center of Military History – Southern France / Rhineland Campaigns
The National Archives

Charles DeGlopper
Congressional Medal of Honor Society
U.S. Army Center of Military History – Medal of Honor Citations
National WWII Museum – Normandy Campaign
U.S. Army Center of Military History – Normandy Campaign
The National Archives

John R. Fox
Congressional Medal of Honor Society
U.S. Army Center of Military History – Medal of Honor Citations
National WWII Museum – Italian Campaign
U.S. Army Center of Military History – Mediterranean Theater / Italy
The National Archives

Harold Gonsalves
Congressional Medal of Honor Society
U.S. Army Center of Military History – Medal of Honor Citations
National WWII Museum – Battle of Okinawa
U.S. Marine Corps History Division
The National Archives

Keep Exploring the Heroes!

We hope you enjoyed reading ***Medal of Honor Heroes: World War II***.

The stories in this book are only the beginning. If you were inspired by these true accounts of courage, sacrifice, and leadership, there are more heroes, more history, and more meaningful stories still to discover.

Discover More Good Buddy Books

Want to explore our full collection of books, get sneak peeks of upcoming titles, or just learn more about what we do? Visit our website!

www.goodbuddybooks.com

Find the Entire *Medal of Honor Heroes* Series

If this book sparked your interest in courage, history, and true stories, explore more titles from Good Buddy Books on Amazon.

Medal of Honor Heroes Series Page (Good Buddy Books)

Answers: Test Your Knowledge

1. B
2. A
3. A
4. B
5. C
6. A
7. B
8. A
9. B
10. A
11. Desmond Doss
12. John Basilone
13. Rodger Young
14. John R. Fox
15. Jack Lummus
16. Charles DeGlopper
17. George Sakato
18. Harold Gonsalves
19. The Medal of Honor is awarded for extraordinary bravery above and beyond the call of duty.
20. Okinawa was important because it was close to Japan and could be used as a base for future military operations.
21. It was made up mostly of Japanese-American soldiers and became one of the most highly decorated units in U.S. military history.
22. They were hard to see, strongly protected, and allowed defenders to fire on advancing troops from cover.
23. Answers may include Desmond Doss, Henry Erwin, Jack Lucas, Hershel "Woody" Williams, Richard Bong, Daniel Inouye, George Sakato, Francis Currey, or Audie Murphy.
24. Answers may include Rodger Young, Jack Lummus, Charles DeGlopper, John R. Fox, or Harold Gonsalves.
25. Answers will vary.

Made in the USA
Monee, IL
04 November 2026

67381213R00098